13th Meeting on the Mathematics of Language 2013

(MoL 13)

Sofia, Bulgaria
9 August 2013

ISBN: 978-1-62748-995-9

Printed from e-media with permission by:

Curran Associates, Inc.
57 Morehouse Lane
Red Hook, NY 12571

Some format issues inherent in the e-media version may also appear in this print version.

Printed by Curran Associates, Inc. (2013)

For permission requests, please contact the Association for Computational Linguistics at the address below.

Association for Computational Linguistics
209 N. Eighth Street
Stroudsburg, Pennsylvania 18360

Phone: 1-570-476-8006
Fax: 1-570-476-0860

acl@aclweb.org

Additional copies of this publication are available from:

Curran Associates, Inc.
57 Morehouse Lane
Red Hook, NY 12571 USA
Phone: 845-758-0400
Fax: 845-758-2634
Email: curran@proceedings.com
Web: www.proceedings.com

MoL 13

The 13th Meeting on the Mathematics of Language

Proceedings

August 9, 2013
Sofia, Bulgaria

Introduction

The Mathematics of Language (MoL) special interest group traces its origins to a meeting held in October 1984 at Ann Arbor, Michigan. While MoL is among the oldest SIGs of the ACL, it is the first time that the proceedings are produced by our parent organization. The first volume was published by Benjamins, later ones became special issues of the *Annals of Mathematics and Artificial Intelligence* and *Linguistics and Philosophy*, and for the last three occasions (really six years, since MoL only meets every second year) we relied on the Springer LNCS series. Perhaps the main reason for this aloofness was that the past three decades have brought the ascendancy of statistical methods in computational linguistics, with the formal, grammar-based methods that were the mainstay of mathematical linguistics viewed with increasing suspicion.

To make matters worse, the harsh anti-formal rhetoric of leading linguists relegated important attempts at formalizing Government-Binding and later Minimalist theory to the fringes of syntax. Were it not for phonology and morphology, where the incredibly efficient finite state methods pioneered by Kimmo Koskenniemi managed to bridge the gap between computational practice and linguistic theory, and were it not for the realization that the mathematical approach has no alternative in machine learning, MoL could have easily disappeared from the frontier of research.

The current volume marks a time when we can begin to see the computational and the theoretical linguistics camps together again. The selection of papers, while still strong on phonology (Heinz and Lai, Heinz and Rogers) and morphology (Kornai et al.), extends well to syntax (Hunter and Dyer, Fowlie) and semantics (Clark et al., Fernando). Direct computational concerns such as machine translation (Martzoukos et al.), decoding (Corlett and Penn), and complexity (Berglund et al.) are now clearly seen as belonging to the core focus of the field.

The 10 papers presented in this volume were selected by the Program Committee from 16 submissions. We would like to thank the authors, the members of the Program Committee, and our invited speaker for their contributions to the planning and execution of the workshop, and the ACL conference organizers, especially Aoife Cahill and Qun Liu (workshops), and Roberto Navigli and Jing-Shin Chang (publications) for their significant contributions to the overall management of the workshop and their direction in preparing the publication of the proceedings.

András Kornai and Marco Kuhlmann (editors)
June 2013

Program Chairs:

András Kornai (Hungarian Academy of Sciences, Hungary)
Marco Kuhlmann (Uppsala University, Sweden)

Program Committee:

Ash Asudeh (Carleton University, Canada)
Alexander Clark (King's College, UK)
Annie Foret (University of Rennes 1, France)
Daniel Gildea (University of Rochester, USA)
Gerhard Jäger (University of Tübingen, Germany)
Aravind Joshi (University of Pennsylvania, USA)
Makoto Kanazawa (National Institute of Informatics, Japan)
Greg Kobele (University of Chicago, USA)
Andreas Maletti (University of Stuttgart, Germany)
Carlos Martín-Vide (University Rovira i Virgili, Spain)
Jens Michaelis (Bielefeld University, Germany)
Gerald Penn (University of Toronto, Canada)
Carl Pollard (The Ohio State University, USA)
Jim Rogers (Earlham College, USA)
Giorgio Satta (University of Padua, Italy)
Noah Smith (Carnegie Mellon University, USA)
Ed Stabler (UCLA, USA)
Mark Steedman (Edinburgh University, UK)
Sylvain Salvati (INRIA, France)
Anssi Yli-Jyrä (University of Helsinki, Finland)

Invited Speaker:

Mark Johnson (Macquarie University, Australia)

Table of Contents

Program

Friday, August 9

Session 1

09:00–09:30 *Distributions on Minimalist Grammar Derivations*
Tim Hunter and Chris Dyer

09:30–10:00 *Order and Optionality: Minimalist Grammars with Adjunction*
Meaghan Fowlie

10:00–10:30 *On the Parameterized Complexity of Linear Context-Free Rewriting Systems*
Martin Berglund, Henrik Björklund and Frank Drewes

10:30–11:00 Coffee Break

Session 2

11:00–11:30 *Segmenting Temporal Intervals for Tense and Aspect*
Tim Fernando

11:30–12:00 *The Frobenius Anatomy of Relative Pronouns*
Stephen Clark, Bob Coecke and Mehrnoosh Sadrzadeh

12:00–12:30 *Vowel Harmony and Subsequentiality*
Jeffrey Heinz and Regine Lai

12:30–14:00 Lunch Break

Friday, August 9 (continued)

Session 3

14:00–14:30 *Learning Subregular Classes of Languages with Factored Deterministic Automata*
Jeffrey Heinz and Jim Rogers

14:30–15:00 *Structure Learning in Weighted Languages*
Andras Kornai, Attila Zséder and Gábor Recski

15:00–15:30 *Why Letter Substitution Puzzles are Not Hard to Solve: A Case Study in Entropy and Probabilistic Search-Complexity*
Eric Corlett and Gerald Penn

15:30–16:00 Coffee Break

Session 4

16:00–16:30 *Investigating Connectivity and Consistency Criteria for Phrase Pair Extraction in Statistical Machine Translation*
Spyros Martzoukos, Christophe Costa Florêncio and Christof Monz

16:30–17:30 *Grammars and Topic Models*
Mark Johnson

Distributions on Minimalist Grammar Derivations

Tim Hunter
Department of Linguistics
Cornell University
159 Central Ave., Ithaca, NY, 14853
tim.hunter@cornell.edu

Chris Dyer
School of Computer Science
Carnegie Mellon University
5000 Forbes Ave., Pittsburgh, PA, 15213
cdyer@cs.cmu.edu

Abstract

We present three ways of inducing probability distributions on derivation trees produced by Minimalist Grammars, and give their maximum likelihood estimators. We argue that a parameterization based on locally normalized log-linear models balances competing requirements for modeling expressiveness and computational tractability.

1 Introduction

Grammars that define not just sets of trees or strings but probability distributions over these objects have many uses both in natural language processing and in psycholinguistic models of such tasks as sentence processing and grammar acquisition. Minimalist Grammars (MGs) (Stabler, 1997) provide a computationally explicit formalism that incorporates the basic elements of one of the most common modern frameworks adopted by theoretical syntacticians, but these grammars have not often been put to use in probabilistic settings. In the few cases where they have (e.g. Hale (2006)), distributions over MG derivations have been over-parametrized in a manner that follows straightforwardly from a conceptualization of the derivation trees as those generated by a particular context-free grammar, but which does not respect the characteristic perspective of the underlying MG derivation. We propose an alternative approach with a smaller number of parameters that are straightforwardly interpretable in terms that relate to the theoretical primitives of the MG formalism. This improved parametrization opens up new possibilities for probabilistically-based empirical evaluation of MGs as a cognitive hypothesis about the discrete primitives of natural language grammars, and for the use of MGs in applied natural language processing.

In Section 2 we present MGs and their equivalence to MCFGs, which provides a context-free characterization of MG derivation trees. We demonstrate the problems with the straightforward method of supplementing a MG with probabilities that this equivalence permits in Section 3, and then introduce our proposed reparametrization that solves these problems in Section 4. Section 5 concludes and outlines some suggestions for future related work.

2 Minimalist Grammars and Multiple Context-Free Grammars

2.1 Minimalist Grammars

A Minimalist Grammar (MG) (Stabler and Keenan, 2003)[1] is a five-tuple $G = \langle \Sigma, Sel, Lic, Lex, \mathrm{c} \rangle$ where:

- Σ is a finite alphabet
- Sel ("selecting types") and Lic ("licensing types") are disjoint finite sets which together determine the set Syn ("syntactic features"), which is the union of the following four sets:

$$\begin{aligned} selectors &= \{=f \mid f \in Sel\} \\ selectees &= \{\ \ f \mid f \in Sel\} \\ licensors &= \{+f \mid f \in Lic\} \\ licensees &= \{-f \mid f \in Lic\} \end{aligned}$$

- Lex ("the lexicon") is a finite subset of $\Sigma^* \times (selectors \cup licensors)^* \times selectees \times licensees^*$
- $\mathrm{c} \in Sel$ is a designated type of completed expressions

(A sample lexicon is shown in Fig. 3 below.)

[1] We restrict attention here to MGs without head movement as presented by Stabler and Keenan (2003). Weak generative capacity is unaffected by this choice (Stabler, 2001).

Given an MG G, an **expression** is an ordered binary tree with non-leaf nodes labeled by an element of $\{<, >\}$, and with leaf nodes labeled by an element of $\Sigma^* \times Syn^*$. We take elements of Lex to be one-node trees, hence expressions. We often write elements of $\Sigma^* \times Syn^*$ with the two components separated by a colon (e.g. *arrive* : +d v). Each application of one of the derivational operations MERGE and MOVE, defined below, "checks" or deletes syntactic features on the expression(s) to which it applies.

The **head** of a one-node expression is the expression's single node; the head of an expression $[_< e_1\ e_2]$ is the head of e_1; the head of an expression $[_> e_1\ e_2]$ is the head of e_2. An expression is **complete** iff the only syntactic feature on its head is a selectee feature c and there are no syntactic features on any of its other nodes. Given an expression e, $yield(e) \in \Sigma^*$ is result of concatenating the leaves of e in order, discarding all syntactic features.

$CL(G)$ is the set of expressions generated by taking the closure of Lex under the functions MERGE and MOVE, defined in Fig. 1; intuitive graphical illustrations are given in Fig 2. The language generated by G is $\{s \mid \exists e \in CL(G)$ such that e is complete and $yield(e) = s\}$.

An example derivation, using the grammar in Fig. 3, is shown in Fig. 4. This shows both the "history" of derivational operations — although operations are not shown explicitly, all binary-branching nodes correspond to applications of MERGE and all unary-branching nodes to MOVE — and the expression that results from each operation. Writing instead only MERGE or MOVE at each internal node would suffice to determine the eventual derived expression, since these operations are functions. A **derivation tree** is a tree that uses this less redundant labeling: more precisely, a derivation tree is either (i) a lexical item, or (ii) a tree $[_{\text{MERGE}}\ \tau_1\ \tau_2]$ such that MERGE$(eval(\tau_1), eval(\tau_2))$ is defined, or (iii) a tree $[_{\text{MOVE}}\ \tau]$ such that MOVE$(eval(\tau))$ is defined; where *eval* is the "interpretation" function that maps a derivation tree to an expression in the obvious way. We define $\Omega(G)$ to be the set of all derivation trees using the MG G.

An important property of the definition of MOVE is that it is only defined on $\tau[+f\alpha]$ if there is a *unique* subtree of this tree whose (head's) first feature is $-f$. From this it follows that in any

pierre : d	*who* : d -wh
marie : d	*will* : =v =d t
praise : =d v	ϵ : =t c
often : =v v	ϵ : =t +wh c

Figure 3: A Minimalist Grammar lexicon. The type of completed expressions is c.

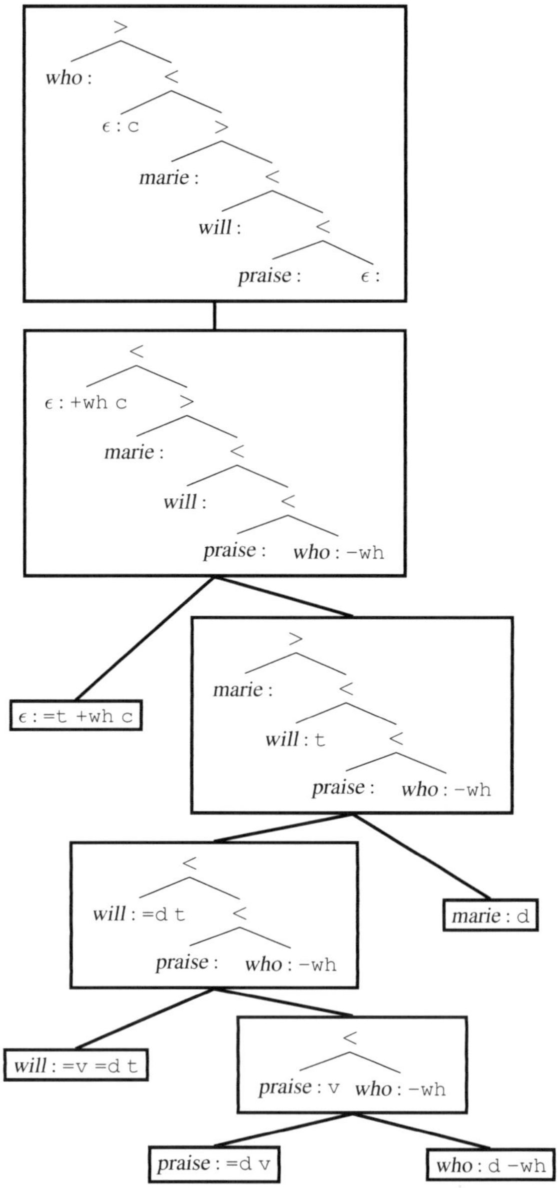

Figure 4: An MG derivation of an embedded question

$$\text{MERGE}(e_1[{=}f\ \alpha], e_2[f\ \beta]) = \begin{cases} [_< e_1[\alpha]\ e_2[\beta]] & \text{if } e_1[{=}f\ \alpha] \in Lex \\ [_> e_2[\beta]\ e_1[\alpha]] & \text{otherwise} \end{cases}$$

$$\text{MOVE}(e_1[{+}f\ \alpha]) = [_> e_2[\beta]\ e_1'[\alpha]]$$

where $e_2[-f\ \beta]$ is a unique subtree of $e_1[+f\ \alpha]$
and e_1' is like e_1 but with $e_2[-f\ \beta]$ replaced by an empty leaf node $\epsilon : \epsilon$

Figure 1: Definitions of MG operations MERGE and MOVE. The first case of MERGE creates complements, the second specifiers. f ranges over $Sel \cup Lic$; α and β range over Syn^*; and $e[\alpha]$ is an MG expression whose head bears the feature-sequence α.

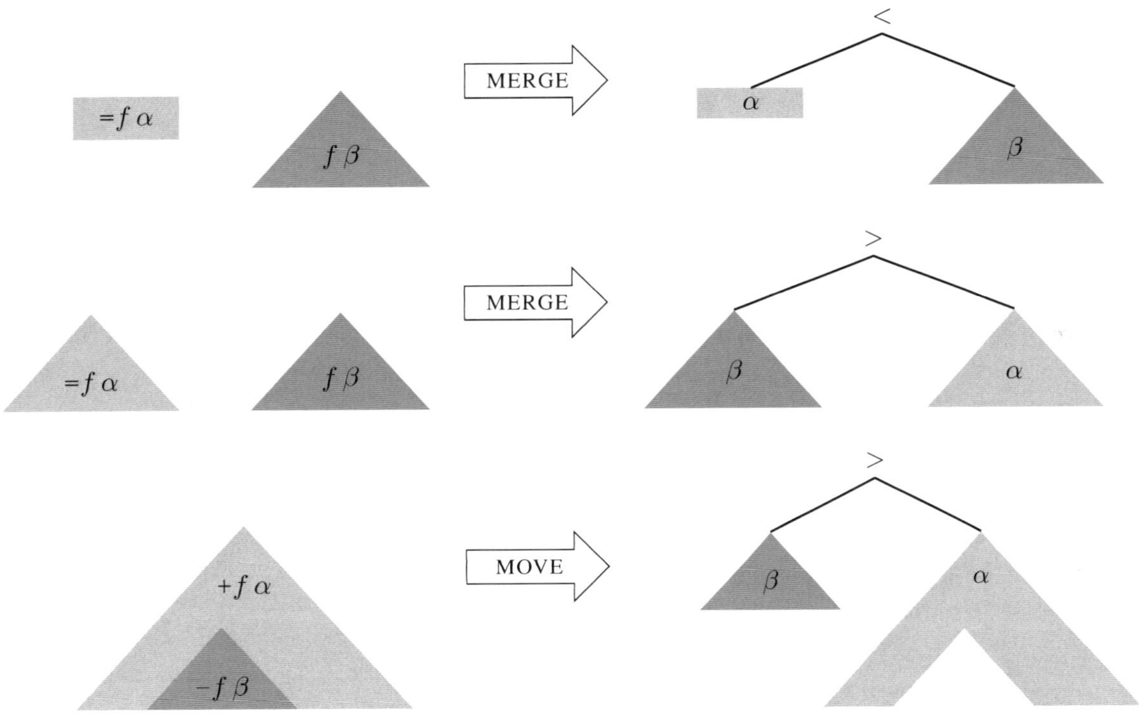

Figure 2: Graphical illustrations of definitions of MERGE and MOVE. Rectangles represent single-node trees. Triangles represent either single-node trees or complex trees, but the second case of MERGE applies only when the first case does not (i.e. when the $=f\ \alpha$ tree is complex).

derivation of a complete expression, every intermediate derived expression will have at most $|Lic|$ subtrees whose (head's) first feature is of the form $-g$ for any $g \in Lic$.

2.2 Multiple Context-Free Grammars

Multiple Context-Free Grammars (MCFGs) (Seki et al., 1991; Kallmeyer, 2010) are a mildly context-sensitive grammar formalism in the sense of Joshi (1985).[2] They bring additional expressive capacity over context-free grammars (CFGs) by generalizing to allow nonterminals to categorize not just single strings, but tuples of strings. For example, while a CFG might categorize *eats cake* as a VP and *the boy* as an NP, an MCFG could categorize the tuple ⟨*says is tall*, *which girl*⟩ as a VPWH (intuitively, a VP containing a WH which will move out of it). Correspondingly, MCFG production rules (construed as recipes for building expressions bottom-up) can specify not only, for example, how to combine a *string* which is an NP and a *string* which is a VP, but also how to combine a *string* which is an NP with a *tuple of strings* which is a VPWH. The CFG rule which would usually be written 'S → NP VP' is shown in (1) in a format that makes explicit the string-concatenation operation; (2) uses this notation to express an MCFG rule that combines an NP with a VPWH to form a string of category Q, an embedded question. (We often omit angle brackets around one-tuples.) An example application of this rule is shown in (3).

$$st :: \text{S} \quad \Rightarrow \quad s :: \text{NP} \qquad t :: \text{VP} \tag{1}$$

$$t_2 s t_1 :: \text{Q} \quad \Rightarrow \quad s :: \text{NP} \quad \langle t_1, t_2\rangle :: \text{VPWH} \tag{2}$$

$$\begin{array}{l} \textit{which girl the boy says is tall} :: \text{Q} \quad \Rightarrow \\ \textit{the boy} :: \text{NP} \quad \langle \textit{says is tall}, \textit{which girl}\rangle :: \text{VPWH} \end{array} \tag{3}$$

Every nonterminal in an MCFG derives (only) n-tuples of strings, for some n known as the nonterminal's **rank**. In the examples above NP, VP, S and Q are of rank 1, and VPWH is of rank 2. A CFG is an MCFG where every nonterminal has rank 1.

Michaelis (2001) showed that it is possible to reformulate MGs in a way that uses categorized string-tuples, of the sort that MCFGs manipulate, as derived structures (or expressions) instead of trees. The "purpose" of the internal tree structure that we assign to derived objects is, in effect, to allow a future application of MOVE to break them apart and rearrange their pieces, as illustrated in Fig. 2. But since the placement of the syntactic features on a tree determines the parts that will be rearranged by a future application of MOVE (in any derivation of a complete expression), we lose no relevant information by splitting up a tree's yield into the components that will be rearranged and then ignoring all other internal structure. Thus the following tree:

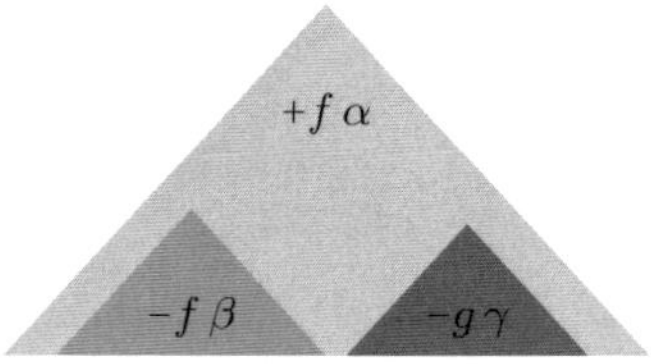

(4)

becomes a tuple of categorized strings (we will explain the 0 subscript shortly):

$$\langle\ s : +f\,\alpha\ ,\ t : -f\,\beta\ ,\ u : -g\,\gamma\ \rangle_0$$

or, equivalently, a tuple of strings, categorized by a tuple-of-categories:

$$\langle s, t, u\rangle :: \langle +f\,\alpha, -f\,\beta, -g\,\gamma\rangle_0 \tag{5}$$

The order of the components is irrelevant *except for* the first component, which contains the entire structure's head node; intuitively, this is the component out of which the others move.

Based on this idea, Michaelis (2001) shows how to construct, for any MG, a corresponding MCFG whose nonterminals are tuples like $\langle +f\,\alpha, -f\,\beta, -g\,\gamma\rangle_0$ from above. The uniqueness requirement in the definition of MOVE ensures that we need only a finite number of such nonterminals. The feature sequences that comprise the MCFG nonterminals, in combination with the MG operations, determine the MCFG production rules in which each MCFG nonterminal appears. For example, the arrangement of features on the tree in (4) dictates that MOVE is the only MG operation that can apply to it; thus the internals of the complex category in (5) correspondingly dictate that the only MCFG production that takes (5) as "input" (again, thinking right-to-left or bottom-up as in (1) and (2)) is one that transforms it in accord with the effects of MOVE. If $\beta = \epsilon$, then this

[2]MCFGs are almost identical to Linear Context-Free Rewrite Systems (Vijay-Shanker et al., 1987). Seki et al. (1991) show that the two formalisms are weakly equivalent.

effect will be to transform the three-tuple into a two-tuple as shown in (6), since the t-component now has no remaining features and has therefore reached its final position:

$$\langle ts, u\rangle :: \langle \alpha, -g\,\gamma\rangle_0 \quad \Rightarrow \quad \langle s, t, u\rangle :: \langle +f\,\alpha, -f, -g\,\gamma\rangle_0 \quad (6)$$

This is analogous — modulo the presence of the additional $u : -g\,\gamma$ component — to the rule that is used in the final step of the derivation in Fig. 5, which is the MCFG equivalent of Fig. 4.

If, on the other hand, $\beta \neq \epsilon$, then the t-component will need to move again later in the derivation, and so we keep it as a separated component:

$$\langle s, t, u\rangle :: \langle \alpha, \beta, -g\,\gamma\rangle_0 \quad \Rightarrow \quad \langle s, t, u\rangle :: \langle +f\,\alpha, -f\,\beta, -g\,\gamma\rangle_0 \quad (7)$$

The subscript 0 on the tuples above indicates that the corresponding expressions are non-lexical; for lexical expressions, the subscript is 1. This information is not relevant to MOVE operations, but is crucial for distinguishing between the complement and specifier cases of MERGE. For example, in the simplest cases where no to-be-moved subconstituents are present, the constructed MCFG must contain two rules corresponding to MERGE as follows. (n matches either 1 or 0.)

$$st :: \langle \alpha\rangle_0 \quad \Rightarrow \quad s :: \langle =f\,\alpha\rangle_1 \quad t :: \langle f\rangle_n \quad (8)$$
$$ts :: \langle \alpha\rangle_0 \quad \Rightarrow \quad s :: \langle =f\,\alpha\rangle_0 \quad t :: \langle f\rangle_n \quad (9)$$

By similar logic, it is possible to construct all the necessary MCFG rules corresponding to MERGE and MOVE; see, for example, Stabler and Keenan (2003, p.347) for (a presentation of the MG operations that can also be straightforwardly be read as) the general schemas that generate these rules. One straightforward lexical/preterminal rule is added for each lexical item in the MG, and the MCFG's start symbol is $\langle \mathtt{c}\rangle_0$.[3] The resulting MCFG is weakly equivalent to the original MG, and strongly equivalent in the sense that one can straightforwardly convert back and forth between the two grammars' derivation trees. The MCFG equivalent of the MG in Fig. 3 is shown in Fig. 6 (ignoring the weights for now, which we come to below).[4]

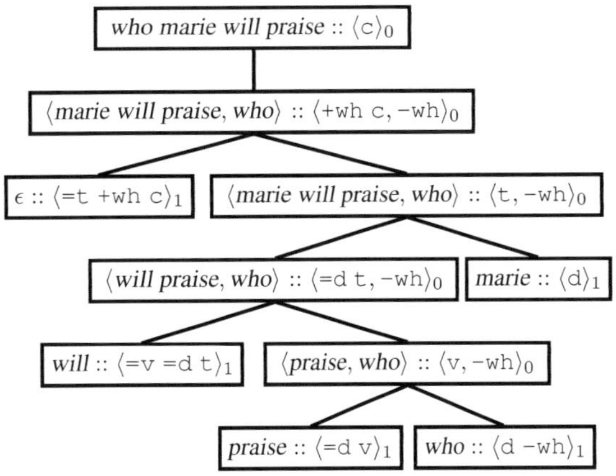

Figure 5: The MG derivation from Fig. 4 illustrated with tuples of strings instead of trees as the derived structures.

Notation. We define the above conversion process to be an (invertible) function π from MGs to MCFGs. That is, for an valid MG, G it holds that $\pi(G)$ is an equivalent MCFG and $\pi^{-1}(\pi(G)) = G$. By abuse of notation, we will use π as the function for converting from MG derivation trees to equivalent MCFG derivation trees. By an MCFG derivation tree we mean a tree like Fig. 5 but with non-leaf nodes labelled only by nonterminals (not tuples of strings). The derivation tree language of an MCFG is thus a local tree language, just as for a CFG; that of an MG is non-local but regular (Kobele et al., 2007).

3 Distributions on Derivations

Assume a Minimalist Grammar, G. In this section and the next, we will consider various ways of defining probability distributions on the derivation trees in $\Omega(G)$.[5] The first approach, introduced in Section 3.2, is conceptually straightforward but is problematic in certain respects that we discuss in Section 3.3. We present a different approach that resolves these problems in Section 4.

We also consider the problem of estimating the parameters of these distributions from a *finite sample* of training data, specified by a function $\tilde{f} : \Omega(G) \to \mathbb{N}$, where $\tilde{f}(\tau)$ is the number of times derivation τ occurs in the sample. To this end, it

[3] We exclude $\langle \mathtt{c}\rangle_1$ on the simplifying assumption that the MG has no lexical item whose only feature is the selectee $\mathtt{c}$.

[4] This MCFG includes only the rules that are "reachable" from the lexical items. For example, we leave aside rules involving the nonterminal $\langle =\mathtt{c}\ \mathtt{v}\ -\mathtt{wh}\rangle_0$, even though the schemas in Stabler and Keenan (2003) generate them.

[5] We use the terms *derivation tree* and *derivation* interchangeably.

θ^{ERF}	
2/2	$\epsilon :: \langle \texttt{=t +wh c} \rangle_1$
95/95	$\epsilon :: \langle \texttt{=t c} \rangle_1$
97/97	*will* $:: \langle \texttt{=v =d t} \rangle_1$
6/6	*often* $:: \langle \texttt{=v v} \rangle_1$
97/97	*praise* $:: \langle \texttt{=d v} \rangle_1$
95/192	*marie* $:: \langle \texttt{d} \rangle_1$
97/192	*pierre* $:: \langle \texttt{d} \rangle_1$
2/2	*who* $:: \langle \texttt{d -wh} \rangle_1$

θ^{ERF}				
2/2	$\langle st, u \rangle :: \langle \texttt{+wh c}, \texttt{-wh} \rangle_0$	$\Rightarrow$	$s :: \langle \texttt{=t +wh c} \rangle_1$	$\langle t, u \rangle :: \langle \texttt{t}, \texttt{-wh} \rangle_0$
95/95	$st :: \langle \texttt{=d t} \rangle_0$	$\Rightarrow$	$s :: \langle \texttt{=v =d t} \rangle_1$	$t :: \langle \texttt{v} \rangle_0$
2/2	$\langle st, u \rangle :: \langle \texttt{=d t}, \texttt{-wh} \rangle_0$	$\Rightarrow$	$s :: \langle \texttt{=v =d t} \rangle_1$	$\langle t, u \rangle :: \langle \texttt{v}, \texttt{-wh} \rangle_0$
2/97	$ts :: \langle \texttt{c} \rangle_0$	$\Rightarrow$	$\langle s, t \rangle :: \langle \texttt{+wh c}, \texttt{-wh} \rangle_0$	
95/97	$st :: \langle \texttt{c} \rangle_0$	$\Rightarrow$	$s :: \langle \texttt{=t c} \rangle_1$	$t :: \langle \texttt{t} \rangle_0$
95/95	$ts :: \langle \texttt{t} \rangle_0$	$\Rightarrow$	$s :: \langle \texttt{=d t} \rangle_0$	$t :: \langle \texttt{d} \rangle_1$
2/2	$\langle ts, u \rangle :: \langle \texttt{t}, \texttt{-wh} \rangle_0$	$\Rightarrow$	$\langle s, u \rangle :: \langle \texttt{=d t}, \texttt{-wh} \rangle_0$	$t :: \langle \texttt{d} \rangle_1$
95/100	$st :: \langle \texttt{v} \rangle_0$	$\Rightarrow$	$s :: \langle \texttt{=d v} \rangle_1$	$t :: \langle \texttt{d} \rangle_1$
5/100	$st :: \langle \texttt{v} \rangle_0$	$\Rightarrow$	$s :: \langle \texttt{=v v} \rangle_1$	$t :: \langle \texttt{v} \rangle_0$
2/3	$\langle s, t \rangle :: \langle \texttt{v}, \texttt{-wh} \rangle_0$	$\Rightarrow$	$s :: \langle \texttt{=d v} \rangle_1$	$t :: \langle \texttt{d -wh} \rangle_1$
1/3	$\langle st, u \rangle :: \langle \texttt{v}, \texttt{-wh} \rangle_0$	$\Rightarrow$	$s :: \langle \texttt{=v v} \rangle_1$	$\langle t, u \rangle :: \langle \texttt{v}, \texttt{-wh} \rangle_0$

Figure 6: The MCFG produced from the MG in Fig. 3, as described in Section 2.2; with weights computed by relative frequency estimation based on the naive parametrization, as described in Section 3.

will be useful to define the **empirical distribution** on derivations to be $\tilde{p}(\tau) = \tilde{f}(\tau)/\sum_{\tau'} \tilde{f}(\tau')$.

3.1 Stochastic MCFGs

As with CFGs, it is straightforward to imbue an MCFG, H, with production probabilities and thereby create a **stochastic MCFG**.[6] In stochastic MCFGs (as in CFGs) the probability of a non-terminal rewrite in a derivation is conditionally independent of all other rewrite decisions, given the non-terminal type. This formulation defines a distribution over MCFG derivations in terms of a random branching process that begins with probability 1 at the start symbol and recursively expands frontier nodes N, drawing branching decisions from the the conditional distribution $p(\cdot \mid N)$; the process terminates when lexical items have been produced on all frontiers.

If $p(\delta \mid N)$ is the probability that N rewrites as δ and $f_\tau(N \Rightarrow \delta)$ is the number of times $N \Rightarrow \delta$ occurs in derivation tree τ, then

$$p(\tau) = \prod_{(N \Rightarrow \delta) \in H} p(\delta \mid N)^{f_\tau(N \Rightarrow \delta)}. \quad (10)$$

With mild assumptions to ensure consistency (Chi, 1999), the $p(\tau)$'s form a proper probability distribution over all derivations in H.[7]

Because the derivation trees of the MG G stand in a bijection with the derivation trees of the MCFG $\pi(G)$, stochastic MCFGs can be used to define a distribution on MG derivations.

[6] Although MCFGs have a greater generative capacity than CFGs, the statistical properties do not change at all, unless otherwise noted.

[7] The estimators that are based on empirical frequencies in a derivation bank which we use in this paper will always yield consistent estimates. Refer Chi (1999) for more detail.

3.2 The naive parametrization

The most straightforward way to parameterize a stochastic MCFG uses individual parameters $\theta_{\delta|N}$ to represent each production probability, i.e., $p(\delta \mid N) \doteq \theta_{\delta|N}$. When applied to an MCFG that is derived from an MG, we will refer to this as the **naive parametrization**.

This is the parametrization used by Hale (2006) to define a probability distribution over the derivations of MGs in order to explore the predictions of an information-theoretic hypothesis concerning sentence comprehension difficulty.

MLE. The arguably most standard technique for setting the parameters of a probability distribution is so that they maximize the likelihood of a sample of training data. In the naive parameterization, the maximum likelihood estimate (MLE) for each parameter $\hat{\theta}^{ERF}_{\delta|N}$ is the *empirical relative frequency* of the rewrite $N \Rightarrow \delta$ in the training data (Abney, 1997):

$$\hat{\theta}^{ERF}_{\delta|N} = \frac{\sum_\tau \tilde{f}(\tau) f_{\pi(\tau)}(N \Rightarrow \delta)}{\sum_\tau \tilde{f}(\tau) \sum_{(N \Rightarrow \delta') \in \pi(G)} f_{\pi(\tau)}(N \Rightarrow \delta')}.$$

3.3 Unfaithfulness to MGs

While the naive parameterization with MLE estimation is simple, it is arguably a poor choice for parameterizing distributions on MGs. The problem is that, relative to the independence assumptions encoded in the MG formalism, each step of the MCFG derivation both conditions on and predicts "too much" structure. As a result, commonalities across different applications of the same MG operation are modeled independently and do not share statistical strength. This arises because

90	pierre will praise marie
5	pierre will often praise marie
1	who pierre will praise
1	who pierre will often praise

Figure 7: An artificial corpus of sentences derivable from the grammars in Figures 3 and 6.

of the way the MCFG's nonterminals multiply out all relevant arrangements of features.[8] We illustrate the problem with an example.

Consider the corpus in Fig. 7, where each sentence is preceded by its frequency. Since each sentence is assigned a unique derivation by our example MG, this is equivalent to a treebank.

One reasonable statistical interpretation of the first two lines is that a verb phrase comprises a verb and an object 95% of the time, and comprises the adverb *often* and another verb phrase 5% of the time (since *pierre will often praise marie* has two nested verb phrase constituents). The last two lines provide an analogous pair of sentences involving wh-movement of the object. A priori, one would expect that the 95:5 relative frequency that describes the presence of the adverb also applies here; however, the ERF estimator will use 2:1 instead. Why is this? The VP category in the MCFG is "split" into two to indicate whether it has a wh-feature inside it, and each has its own parameters. We criticize this on the grounds that it is not in line with our main goal of defining a distribution over the derivations *of the MG*: from the perspective of the MG, there is a sense in which it is "the same instance" of MERGE that combines *often* with a verb phrase, whether or not the verb phrase's object bears a -wh feature. In other words, the differences between the following two trees seem unrelated to the way in which they are both candidates to be merged with *often* : =v v.

From the perspective of the MCFG, however, the introduction of the adverb is mediated by expansions of the nonterminal $\langle \mathrm{v} \rangle_0$ in cases without object wh-movement, but by expansions of the distinct nonterminal $\langle \mathrm{v}, \text{-wh} \rangle_0$ in cases with it. Therefore the information about adverb inclusion that is conveyed by the movement-free entries in the corpus is interpreted as only relevant to similarly movement-free derivations. This can be seen in the weights of the last four rules in Fig. 6, which were computed by relative frequency estimation on the basis of the corpus.

Relative to the underlying MG, the naive parametrization has too many degrees of freedom: the model is *overparameterized* and is capable of capturing statistical distinctions that we have theoretical reasons to dislike. Of course, it is possible that VPs have meaningfully different distributions depending on whether or not they contain a wh-feature; however, we would like a parameterization that provides the flexibility to treat these two different contexts as identical, as different, or to share statistical strength between them in some other way. In the next section we propose two alternative parametrizations that provide this control.

4 Log-linear MCFGs

4.1 Globally normalized log-linear models

An alternative mechanism for inducing a distribution on $\Omega(G)$ that provides more control over independence assumptions is the **globally normalized log-linear model** (also called a Markov random field, undirected model, or Gibbs distribution). Unlike the model in the previous section, log-linear models are *not* stochastic in nature—they assign probabilities to structured objects, but they do not rely on a random branching process to do so. Rather, they use a d-dimensional vector of feature functions $\mathbf{\Phi} = \langle \Phi_1, \Phi_2, \ldots, \Phi_d \rangle$, where $\Phi_i : \Omega(G) \to \mathbb{R}$, to extract *features* of the derivation, and a real-valued weight vector $\boldsymbol{\lambda} \in \mathbb{R}^d$.[9] Together, $\mathbf{\Phi}$ and $\boldsymbol{\lambda}$ define the *score* of a derivation τ as a monotonic function of the weighted sum of the feature values $\Phi_1(\tau), \ldots, \Phi_d(\tau)$:

$$s_{\boldsymbol{\lambda}}(\tau) = \exp(\boldsymbol{\lambda} \cdot \mathbf{\Phi}(\tau)).$$

Using this function, a **Gibbs distribution** on the derivations in $\Omega(G)$ is

$$p_{\boldsymbol{\lambda}}(\tau) = \frac{s_{\boldsymbol{\lambda}}(\tau)}{\sum_{\tau' \in \Omega(G)} s_{\boldsymbol{\lambda}}(\tau')}, \qquad (11)$$

[8] Stabler (forthcoming) also discusses the sense in which MCFG rules "miss generalizations" found in MGs.

[9] The term **feature** here refers to functions of a derivation; it should *not* be confused with the **syntactic features** discussed immediately above. However, in as much as syntactic features characterize the steps in a derivation, it is natural that they would play a central role in defining distributions over derivations, and indeed, our proposed feature functions examine syntactic features almost exclusively.

provided that the sum in the denominator is finite.[10]

Notice that (11) is similar to the formula for a relative frequency, the difference being that we use a derivation's score $s_{\boldsymbol{\lambda}}(\tau)$ rather than its empirical count. This use of scores provides a way to express the kind of "missed similarities" we discussed in Section 3.3 via the choice of feature functions. Returning to the example from above, in order to express the similarity between the two adverb-introducing rules — one involving the nonterminal $\langle \mathtt{v} \rangle_0$, the other involving $\langle \mathtt{v}, \mathtt{-wh} \rangle_0$ — we could define a particular feature function Φ_i that maps a derivation to 1 if it contains either one of these rules and 0 otherwise. Then, all else being equal, setting the corresponding parameter λ_i to a higher value will increase the score $s_{\boldsymbol{\lambda}}(\tau)$, and hence the probability $p_{\boldsymbol{\lambda}}(\tau)$, of *any* derivation τ that introduces an adverb, with or without wh-movement of the object.

MLE. As with the naive parameterization, the the parameters $\boldsymbol{\lambda}$ may be set to maximize the (log) likelihood of the training data, i.e.,

$$\hat{\boldsymbol{\lambda}} = \arg\max_{\boldsymbol{\lambda}} \prod_{i=1}^{n} p_{\boldsymbol{\lambda}}(\tau_i)^{\tilde{f}(\tau_i)}$$
$$= \arg\max_{\boldsymbol{\lambda}} \underbrace{\sum_{i=1}^{n} \tilde{f}(\tau_i) \log p_{\boldsymbol{\lambda}}(\tau_i)}_{=\mathcal{L}\ \text{[log likelihood]}} . \qquad (12)$$

We remark that maximizing the log likelihood of data in this parameterization is equivalent to finding the distribution $p_{\boldsymbol{\lambda}}(\tau)$ in which the expected value of $\boldsymbol{\Phi}(\tau)$ is equal to the expected value of the same under the empirical distribution (i.e., under $\tilde{p}(\tau)$) and whose *entropy is maximized* (Della Pietra et al., 1997). This equivalence is particularly clear when the gradient of $\mathcal{L}$ (see (12)) with respect to $\boldsymbol{\lambda}$ is examined:

$$\nabla_{\boldsymbol{\lambda}} \mathcal{L} = \mathbb{E}_{\tilde{p}(\tau)}[\boldsymbol{\Phi}(\tau)] - \mathbb{E}_{p_{\boldsymbol{\lambda}}(\tau)}[\boldsymbol{\Phi}(\tau)]. \qquad (13)$$

This form makes clear that $\mathcal{L}$ achieves an optimum when the expectations of $\boldsymbol{\Phi}$ match under the two distributions.[11]

4.2 Feature locality

Notice that the approach just outlined is extremely general: the feature functions $\boldsymbol{\Phi}$ can examine the derivation trees as a whole. It is possible to define features that pay attention to arbitrary or global properties of a derivation. While such features might in fact generalize well to new data — for example, one could mimic a bigram language model by including features counting bigrams in the string that is generated by the derivation — these are intuitively "bad" since they ignore the derivation's structure. Furthermore, there is a substantial practical downside to allowing unrestricted feature definitions: features that do not "agree" with the derivation structure make inference computationally intractable. Specifically, finding the best most probable derivation of a sentence with "global" features is NP-hard (Koller and Friedman, 2009).

For these reasons, it is advantageous to require that $\boldsymbol{\Phi}$ *decompose additively* in terms of *local* feature functions, φ over the steps that make up a derivation. For defining distributions under an MG G, we will assume that feature functions decompose over the productions in a derivation under the MCFG projection $\pi(G)$, i.e.,

$$\boldsymbol{\Phi}(\tau) = \sum_{(N \Rightarrow \delta) \in \pi(\tau)} \varphi(N \Rightarrow \delta) .$$

Under the locality assumption, we may rewrite the score $s_{\boldsymbol{\lambda}}(\tau)$ as

$$\prod_{(N \Rightarrow \delta) \in \pi(G)} (\exp(\boldsymbol{\lambda} \cdot \varphi(N \Rightarrow \delta)))^{f_{\pi(\tau)}(N \Rightarrow \delta)} .$$

This (partially) addresses the issue of computational tractability, enforces our intuition that the score of a derivation tree should be a function of scores of its component steps, and still gives us the ability to avoid the overconditioning that we identified in Section 3.3.[12]

4.3 Locally normalized log-linear models

Even with our assumption of feature locality, finding $\hat{\boldsymbol{\lambda}}$ remains challenging since the second term

[10]There are several conditions under which this is true. It is trivially true if $|\Omega(G)| < \infty$. When Ω is infinite, the denominator may still be finite if features functions grow (super) linearly with the derivation size in the limiting case as the size tends to infinity. Then, if feature weights are negative, the denominator will either be equal to or bounded from above by an infinite geometric series with a finite sum. Refer to Goodman (1999) and references therein.

[11]While the maximizing point cannot generally be solved for analytically, gradient based optimization techniques may be effectively used to find it (and it is both guaranteed to exist and guaranteed to be unique).

[12]We say that the issue of computational tractability is only *partially* resolved because only certain operations — identifying the most probable derivation of a string — are truly efficient. Computing the model's normalization function, while no longer NP-hard, still not practical.

in (13) is difficult to compute.[13] In this section we suggest a parameterization that admits both efficient ML estimation and retains the ability to use feature functions to control the distribution.

To do so, we revisit the approach of defining distributions on derivations in terms of a stochastic process from Section 3.1, but rather than defining the branching distributions with independent parameters for each MCFG nonterminal rewrite type, we parameterize it in terms of locally normalized log-linear models, also called a **conditional logit model** (Murphy, 2012). Given an MG G, a weight vector $\boldsymbol{w} \in \mathbb{R}^d$, and rule-local feature functions $\boldsymbol{\varphi}$ as defined above,[14] let the branching probability

$$p_{\boldsymbol{w}}(\delta \mid N) \doteq \frac{\exp(\boldsymbol{w} \cdot \boldsymbol{\varphi}(N \Rightarrow \delta))}{\sum_{(N \Rightarrow \delta') \in \pi(G)} \exp(\boldsymbol{w} \cdot \boldsymbol{\varphi}(N \Rightarrow \delta'))}.$$

Like the parametrization in Section 4.1, this new parametrization is based on log-linear models and therefore allows us to express similarities among derivational operations via choices of feature functions. However, rather than defining feature functions Φ_i on entire derivations, these features can only "see" individual MCFG rules. Put differently, the same technique we used in Section 4.1 to define a probability distribution over the entire set of derivations, is used here to define each of the local conditional probability distributions over the expansions of a single MCFG nonterminal. Via the perspective familiar from stochastic MCFGs, these individual conditional probability distributions together define a distribution on the entire set of derivations.

MLE. As with the previous two models, we can set parameters $\boldsymbol{w}$ to maximize the likelihood of the training data. Here, the global likelihood is expressed in terms of the probabilities of conditionally independent rewrite events, each defined in a log-linear model:

$$\mathcal{L}^c = \sum_{\tau} \tilde{f}(\tau) \sum_{(N \Rightarrow \delta) \in \pi(\tau)} f_{\pi(\tau)}(N \Rightarrow \delta) \log p_{\boldsymbol{w}}(\delta \mid N).$$

[13] Specifically, it requires computing expectations under all possible derivations in $\Omega(\pi(G))$ during each step of gradient ascent, which requires polynomial space/time in the size of the lexicon to compute exactly.

[14] The notational shift from $\boldsymbol{\lambda}$ to $\boldsymbol{w}$ to emphasizes that these two parameter vectors have very different semantics. The former parameterizes potential functions in a globally normalized random field while the later is used to determine a family of conditional probability distributions used to define a stochastic process.

Its gradient with respect to $\boldsymbol{w}$ is therefore

$$\nabla_{\boldsymbol{w}} \mathcal{L}^c = \sum_{\tau} \tilde{f}(\tau) \sum_{(N \Rightarrow \delta) \in \pi(\tau)} f_{\pi(\tau)}(N \Rightarrow \delta) \Big[\boldsymbol{\varphi}(N \Rightarrow \delta) - \mathbb{E}_{p_{\boldsymbol{w}}(\delta' \mid N)} \boldsymbol{\varphi}(N \Rightarrow \delta') \Big].$$

As with the globally normalized model, $\nabla_{\boldsymbol{w}} \mathcal{L}^c = 0$ has no closed form solution; however, gradient-based optimization is likewise effective. However, unlike (13), this gradient is straightforward to compute since it requires summing only over the different rewrites of each non-terminal category during each iteration of gradient ascent, rather than over all possible derivations in $\Omega(G)$!

4.4 Example parameter estimation

In this section we compare the probability estimates for productions in a stochastic MCFGs obtained using the naive parameterization discussed in Section 3.2 that conditions on "too much" information and those obtained using locally normalized log-linear models with grammar-appropriate feature functions. Our very simple feature set consists just of binary-valued feature functions that indicate:

- whether a MERGE step, MOVE step, or a terminating lexical-insertion step is being generated;
- what selector feature (in the case of MERGE steps) or licensor feature (in the case of MOVE steps) is being checked (e.g., `+wh` or `=d` or `=v`); and
- what lexical item is used (e.g., *marie* : `d` or ϵ : `=t c`), in the case of terminating lexical-insertion steps.

Table 1 shows the values of some of these features for a sample of the MCFG rules in Fig. 6.

Table 2 compares the production probabilities estimated for last four rules in Fig. 6 using the naive empirical frequency method and our recommended log-linear approach with the features defined as above.[15] The presence or absence of a `-wh` feature does not affect the log-linear model's probability of adding an adverb to a verb phrase, in keeping with the perspective suggested by the derivational operations of MGs.

[15] The log-linear parameters were optimized using a standard quasi-Newtonian method (Liu and Nocedal, 1989).

Table 1: Selected feature values for a sample of MCFG rules. The first four rules are the ones that illustrated the problems with the naive parametrization in Section 3.3.

MCFG Rule	φ_{MERGE}	$\varphi_{\texttt{=d}}$	$\varphi_{\texttt{=v}}$	$\varphi_{\texttt{=t}}$	φ_{MOVE}	$\varphi_{\texttt{+wh}}$
$st :: \langle \texttt{v} \rangle_0 \Rightarrow s :: \langle \texttt{=d v} \rangle_1 \quad t :: \langle \texttt{d} \rangle_1$	1	1	0	0	0	0
$st :: \langle \texttt{v} \rangle_0 \Rightarrow s :: \langle \texttt{=v v} \rangle_1 \quad t :: \langle \texttt{v} \rangle_0$	1	0	1	0	0	0
$\langle s, t \rangle :: \langle \texttt{v}, \texttt{-wh} \rangle_0 \Rightarrow s :: \langle \texttt{=d v} \rangle_1 \quad t :: \langle \texttt{d -wh} \rangle_1$	1	1	0	0	0	0
$\langle st, u \rangle :: \langle \texttt{v}, \texttt{-wh} \rangle_0 \Rightarrow s :: \langle \texttt{=v v} \rangle_1 \quad \langle t, u \rangle :: \langle \texttt{v}, \texttt{-wh} \rangle_0$	1	0	1	0	0	0
$st :: \langle \texttt{c} \rangle_0 \Rightarrow s :: \langle \texttt{=t c} \rangle_1 \quad t :: \langle \texttt{t} \rangle_0$	1	0	0	1	0	0
$ts :: \langle \texttt{c} \rangle_0 \Rightarrow \langle s, t \rangle :: \langle \texttt{+wh c}, \texttt{-wh} \rangle_0$	0	0	0	0	1	1

Table 2: Comparison of probability estimators.

MCFG Rule	Naive $\hat{p}$	Log-linear $\hat{p}$
$st :: \langle \texttt{v} \rangle_0 \Rightarrow s :: \langle \texttt{=d v} \rangle_1 \quad t :: \langle \texttt{d} \rangle_1$	0.95	0.94
$st :: \langle \texttt{v} \rangle_0 \Rightarrow s :: \langle \texttt{=v v} \rangle_1 \quad t :: \langle \texttt{v} \rangle_0$	0.05	0.06
$\langle s, t \rangle :: \langle \texttt{v}, \texttt{-wh} \rangle_0 \Rightarrow s :: \langle \texttt{=d v} \rangle_1 \quad t :: \langle \texttt{d -wh} \rangle_1$	0.67	0.94
$\langle st, u \rangle :: \langle \texttt{v}, \texttt{-wh} \rangle_0 \Rightarrow s :: \langle \texttt{=v v} \rangle_1 \quad \langle t, u \rangle :: \langle \texttt{v}, \texttt{-wh} \rangle_0$	0.33	0.06

5 Conclusion and Future Work

We have presented a method for inducing a probability distribution on the derivations of a Minimalist Grammar in a way that remains faithful to the way the derivations are conceived of in this formalism, and for obtaining the maximum likelihood estimate of its parameters. Our proposal takes advantage of the MG-MCFG equivalence in the sense that it uses the underlying probabilistic branching process of a stochastic MCFG, but avoids the problems of overparametrization that come with the naive approach that reifies the MCFG itself.

Our parameterization has several applications worth noting. It provides a new way to compare variants of the MG formalism that propose slightly different sets of primitives (operations, types of features, etc.) but are equivalent once transformed into MCFGs. Examples of such variants include the addition of an ADJOIN operation (Frey and Gärtner, 2002), or replacing MERGE and MOVE with a single feature-checking operation (Stabler, 2006; Hunter, 2011). Derivations using these different versions of the formalism often boil down to the same string-concatenation operations and will therefore be expressible using equivalent sets of MCFG rules. The naive parametrization will therefore not distinguish them, but in the same way that our proposal above "respects" standard MGs' classification of MCFG rules according to one set of derivational primitives, one could define feature vectors that respect different classifications.

Outside of MGs, the strategy is applicable to any other formalisms whose derivations can be recast as those of MCFGs, such as TAGs and CCGs. More generally still, it could be applied to any formalism whose derivation tree languages can be characterized by a local tree grammar; in our case, the relevant local tree language is obtained via a projection from the regular tree language of MG derivation trees.

Acknowledgments

Thanks to John Hale for helpful discussion and to the anonymous reviewers for their insightful comments. This work was sponsored by NSF award number 0741666, and by the U. S. Army Research Laboratory and the U. S. Army Research Office under contract/grant number W911NF-10-1-0533.

References

Steven P. Abney. 1997. Stochastic attribute-value grammars. *Computational Linguistics*.

Zhiyi Chi. 1999. Statistical properties of probabilistic context-free grammars. *Computational Linguistics*.

Stephen Della Pietra, Vincent Della Pietra, and John Lafferty. 1997. Inducing features of random fields. *IEEE Transactions on Pattern Analysis and Machine Intelligence*, 19(4).

Werner Frey and Hans-Martin Gärtner. 2002. On the treatment of scrambling and adjunction in minimalist grammars. In Gerhard Jäger, Paola Monachesi, Gerald Penn, and Shuly Wintner, editors, *Proceedings of Formal Grammar 2002*, pages 41–52.

Joshua Goodman. 1999. Semiring parsing. *Computational Linguistics*, 25(4).

John Hale. 2006. Uncertainty about the rest of the sentence. *Cognitive Science*, 30:643–672.

Tim Hunter. 2011. Insertion Minimalist Grammars: Eliminating redundancies between merge and move. In Makoto Kanazawa, András Kornai, Marcus Kracht, and Hiroyuki Seki, editors, *The Mathematics of Language (MOL 12 Proceedings)*, volume 6878 of *LNCS*, pages 90–107. Springer, Berlin Heidelberg.

Aravind Joshi. 1985. How much context-sensitivity is necessary for characterizing structural descriptions? In David Dowty, Lauri Karttunen, and Arnold Zwicky, editors, *Natural Language Processing: Theoretical, Computational and Psychological Perspectives*, pages 206–250. Cambridge University Press, New York.

Laura Kallmeyer. 2010. *Parsing Beyond Context-Free Grammars*. Springer-Verlag, Berlin Heidelberg.

Gregory M. Kobele, Christian Retoré, and Sylvain Salvati. 2007. An automata theoretic approach to minimalism. In James Rogers and Stephan Kepser, editors, *Proceedings of the Workshop on Model-Theoretic Syntax at 10; ESSLLI '07*.

Daphne Koller and Nir Friedman. 2009. *Probabilistic Graphical Models: Principles and Techniques*. MIT Press.

Dong C. Liu and Jorge Nocedal. 1989. On the limited memory BFGS method for large scale optimization. *Mathematical Programming B*, 45(3):503–528.

Jens Michaelis. 2001. Derivational minimalism is mildly context-sensitive. In Michael Moortgat, editor, *Logical Aspects of Computational Linguistics, LACL 1998*, volume 2014 of *LNCS*, pages 179–198. Springer, Berlin Heidelberg.

Kevin P. Murphy. 2012. *Machine Learning: A Probabilistic Perspective*. MIT Press.

Hiroyuki Seki, Takashi Matsumara, Mamoru Fujii, and Tadao Kasami. 1991. On multiple context-free grammars. *Theoretical Computer Science*, 88:191–229.

Edward P. Stabler and Edward L. Keenan. 2003. Structural similarity within and among languages. *Theoretical Computer Science*, 293:345–363.

Edward P. Stabler. 1997. Derivational minimalism. In Christian Retoré, editor, *Logical Aspects of Computational Linguistics*, volume 1328 of *LNCS*, pages 68–95. Springer, Berlin Heidelberg.

Edward P. Stabler. 2001. Recognizing head movement. In Philippe de Groote, Glyn Morrill, and Christian Retoré, editors, *Logical Aspects of Computational Linguistics*, volume 2099 of *LNCS*, pages 254–260. Springer, Berlin Heidelberg.

Edward P. Stabler. 2006. Sidewards without copying. In Shuly Wintner, editor, *Proceedings of The 11th Conference on Formal Grammar*, pages 157–170. CSLI Publications, Stanford, CA.

Edward Stabler. forthcoming. Two models of minimalist, incremental syntactic analysis. *Topics in Cognitive Science*.

K. Vijay-Shanker, David J. Weir, and Aravind K. Joshi. 1987. Characterizing structural descriptions produced by various grammatical formalisms. In *Proc. 25th Meeting of Assoc. Computational Linguistics*, pages 104–111.

Order and Optionality: Minimalist Grammars with Adjunction

Meaghan Fowlie
UCLA Linguistics
Los Angeles, California
mfowlie@ucla.edu

Abstract

Adjuncts are characteristically optional, but many, such as adverbs and adjectives, are strictly ordered. In Minimalist Grammars (MGs), it is straightforward to account for optionality or ordering, but not both. I present an extension of MGs, MGs with Adjunction, which accounts for optionality and ordering simply by keeping track of two pieces of information at once: the original category of the adjoined-to phrase, and the category of the adjunct most recently adjoined. By imposing a partial order on the categories, the Adjoin operation can require that higher adjuncts precede lower adjuncts, but not vice versa, deriving order.

1 Introduction

The behaviour of adverbs and adjectives has qualities of both ordinary selection and something else, something unique to that of modifiers. This makes them difficult to model. Modifiers are generally optional and transparent to selection while arguments are required and driven by selection. In languages with relatively strict word order, arguments are strictly ordered, while modifiers may or may not be. In particular, (Cinque, 1999) proposes that adverbs, functional heads, and descriptive adjectives are underlyingly uniformly ordered across languages and models them by ordinary Merge or selection. Such a model captures only the ordering restrictions on these morphemes; it fails to capture their apparent optionality and transparency to selection. I propose a model of these ordered yet optional and transparent morphemes that introduces a function Adjoin which operates on pairs of categories: the original category of the modified phrase together with the category of the most recently adjoined modifier. This allows the derivation to keep track of both the true head of the phrase and the place in the Cinque hierarchy of the modifier, preventing inverted modifier orders in the absence of Move.

2 Minimalist Grammars

I formulate my model as a variant of *Minimalist Grammars* (MGs), which are Stabler (1997)'s formalisation of Chomsky's (1995) notion of feature-driven derivations using the functions Merge and Move. MGs are mildly context-sensitive, putting them in the right general class for human language grammars. They are also simple and intuitive to work with. Another useful property is that the properties of well-formed *derivations* are easily separated from the properties of *derived structures* (Kobele et al., 2007). Minimalist Grammars have been proposed in a number of variants, with the same set of well-formed derivations, such as the string-generating grammar in Keenan & Stabler (2003), the tree-generating grammars in Stabler (1997) and Kobele et al (2007), and the multidominant graph-generating grammar in Fowlie (2011).

At the heart of each of these grammars is a function that takes two derived structures and puts them together, such as string concatenation or tree/graph building. To make this presentation as general as possible, I will simply call these functions **Com**. I will give derived structures as strings as (2003)'s grammar would generate them,[1] but this is just a place-holder for any derived structure the grammar might be defined to generate.

Definition 2.1. A ***Minimalist Grammar*** is a five-tuple $G = \langle \Sigma, \mathbf{sel}, \mathbf{lic}, Lex, M \rangle$. Σ is a finite set of symbols called the *alphabet*. **sel**$\cup$**lic** are finite sets of *base features*. Let F={`+f`, `-f`, `=X`, `X` | `f`$\in$

[1] Keenan & Stabler's grammar also incorporates an additional element: lexical items are triples of string, features, and lexical status, which allows derivation of Spec-Head-Complement order. I will leave this out for simplicity, as it is not relevant here.

lic, X $\in$ **sel**$\}$ be the *features*. For ϵ the empty string, $Lex \subseteq \Sigma \cup \{\epsilon\} \times F^*$ is the *lexicon*, and M is the set of operations **Merge** and **Move**. The language L_G is the closure of Lex under M. A set $C \subseteq F$ of designated features can be added; these are the types of complete sentences.

Minimalist Grammars are *feature-driven*, meaning features of lexical items determine which operations can occur and when. There are two disjoint finite sets of features, ***selectional*** features **sel** which drive the operation Merge and ***licensing*** features **lic** which drive Move. Merge puts two derived structures together; Move operates on the already built structure. Each feature has a positive and negative version, and these features with their polarities make the set F from which the feature stacks for Lexical Items are drawn. In the course of the derivation the features will be *checked*, or deleted, by the operations Merge and Move.

Polarity→	Pos	Neg	
for **Merge**	=X	X	X$\in$ **sel**
for **Move**	+f	-f	f$\in$ **lic**

Table 1: Features

In order for a derivation to succeed, LIs must be in the following form:

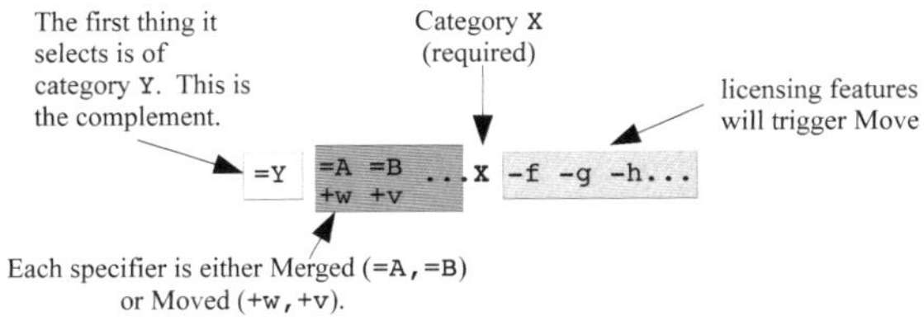

Figure 1: LI template

For example, $\langle$kick, =D=DV$\rangle$ takes a complement of category D, a specifier of category D, and is itself a V. $\langle$which, =ND-wh$\rangle$ takes an N as complement forming a D phrase, which will move because of feature wh.

Merge and Move are defined over ***expressions***: sequences of pairs $\langle$derived structure, feature stack$\rangle$. The first pair in the sequence can be thought of as the "main" structure being built; the remaining are waiting to move. An expression ***displays*** feature f just in case that feature is the first feature in the feature stack of the first pair.

An MG essentially works as follows: Merge is a binary operation driven by **sel**. It takes two expressions and combines them into one just in case the first expression displays =X and the second displays X for some X $\in$ **sel**. Once the second expression is selected, it may still have features remaining; these are always negative licensing features and mean that the second structure is going to move. As such it is stored separately by the derivation. When the matching positive licensing feature comes up later in the derivation, the moving structure is combined again. This is Move.

Move also carries the requirement that for each f$\in$**lic** there be at most one structure waiting to move. This is the ***shortest move constraint (SMC)***.[2]

Definition 2.2 (*Merge*). For α, β sequences of negative **lic** features, s, t derived structures:[3]

$$\textbf{Merge}(\langle s,\ \texttt{=X}\alpha\rangle :: \text{movers}_s,\ \langle t,\ \texttt{X}\beta\rangle :: \text{movers}_t) = \begin{cases} (\textbf{Com}(s,t),\alpha) :: \text{movers}_s \cdot \text{movers}_t & \text{if } \beta = \epsilon \\ (s,\alpha) :: (t,\beta) :: \text{movers}_s \cdot \text{movers}_t & \text{if } \beta \neq \epsilon \end{cases}$$

Definition 2.3 (*Move*). For α, β, γ sequences of negative **lic** features, s, t derived structures, suppose $\exists! \langle t, \beta\rangle \in$ movers such that $\beta = \texttt{-f}\gamma$. Then: $\textbf{Move}(\langle s,\ \texttt{+f}\alpha\rangle$ ::movers$) =$

$$\begin{cases} \langle\textbf{Com}(s,t),\alpha\rangle :: \text{movers} - \langle t,\beta\rangle & \text{if } \gamma = \epsilon \\ \langle s,\alpha\rangle :: \langle t,\gamma\rangle :: \text{movers} - \langle t,\beta\rangle & \text{if } \gamma \neq \epsilon \end{cases}$$

In this article I will make use of *annotated derivation trees*, which are trees describing the derivation. In addition to the name of the function, I (redundantly) include for clarity the derived expressions in the form of strings and features, and sometimes an explanation of why the function applied. For example, Figure 2 shows derivations (unannotated and annotated) of *the wolf* with feature D.

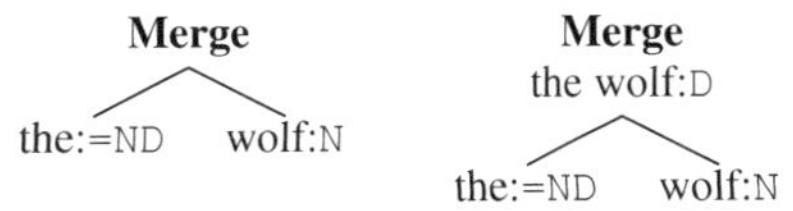

Figure 2: Unannotated and annotated derivation trees

[2]The SMC is based on economy arguments in the linguistic literature (Chomsky, 1995), but it is also crucial for a type of finiteness: the valid derivation trees of an MG form a regular tree language (Kobele et al., 2007). The number of possible movers must be finite for the automaton to be finite-state. The SMC could also be modified to allow up to a particular (finite) number of movers for each f$\in$**lic**.

[3]:: adds an element to a list; $\cdot$ appends two lists; $-$ removes an element from a list.

3 Cartography

The phenomena this model is designed to account for are modifiers and other apparently optional projections such as the following:

(1) a. The small ancient triangular green Irish pagan metal artifact was lost.
b. *The metal green small artifact was lost. **Adjectives**
c. Frankly, John probably once usually arrived early.
d. *Usually, John early frankly once arrived probably. **Adverbs**
e. [$_{DP}$ zhe [$_{NumP}$ yi [$_{ClP}$ zhi [$_{NP}$ bi]]]
[$_{DP}$ this [$_{NumP}$ one [$_{ClP}$ CL [$_{NP}$ pen]]]
'this pen' **Functional projections**

These three phenomena can all display *optionality, transparency to selection*, and *strict ordering*. By transparency I mean that despite the intervening modifiers, properties of the selected head are relevant to selection. For example, in a classifier language, the correct classifier selects a noun even if adjectives intervene.

The hypothesis that despite their optionality these projections are strictly ordered is part of *syntactic cartography* (Rizzi, 2004). Cinque (1999, 2010) in particular proposes a universal hierarchy of functional heads that select adverbs in their specifiers, yielding an order on both the heads and the adverbs. He proposes a parallel hierarchy of adjectives modifying nouns. These hierarchies are very deep. The adverbs and functional heads incorporate 30 heads and 30 adverbs.

Cinque argues that the surprising universality of adverb order calls for explanation. For example, Italian, English, Norwegian, Bosnian/Serbo-Croatian, Mandarin Chinese, and more show strong preferences for *frankly* to precede *(un)fortunately*. These arguments continue for a great deal more adverbs.[4]

(2) Italian
a. **Francamente** ho *purtroppo* una
Frankly have *unfortunately* a
pessima opinione di voi.
bad opinion of you
'Frankly I unfortunately have a very bad opinion of you.'
b. *Purtroppo ho francamente una pessima
Unfortuately have **frankly** a bad
opinione di voi.
opinion of you

(3) English
a. **Frankly**, I *unfortuately* have a very bad opinion of you
b. ?*Unfortunately* I **frankly** have a very bad opinion of you

(4) Norwegian
a. Per forlater [**rerlig talt**] *[heldigvis]*
Peter leaves [**honestly spoken**] *[fortunately]*
[nil] selskapet.
[now] the.party.
'Frankly, Peter is fortunately leaving the party now.'
b. *Per forlater *[heldigvis]* [**rerlig talt**]
Peter leaves *[fortunately]* [**honestly spoken**]
[nil] selskapet.
[now] the.party.

(5) Bosnian/Serbo-Croatian
a. **Iskreno**, ja *naialost* imam jako lose
Frankly, I *unfortunately* have very bad
misljenje o vama
opinion of you.
Frankly, I unfortunately have a very bad opinion of you.'
b. **Naialost*, ja **iskreno** imam jako lose
unfortunately I **frankly** have very bad
misljenje o varna.
opinion of you.

(6) Mandarin Chinese
a. **laoshi-shuo** wo *buxing* dui tamen you
Frankly, I *unfortunately* to them have
pian-jian.
prejudice
'Honestly I unfortunately have prejudice against them.'
b. **buxing* wo **laoshi-shuo** dui tamen you
unfortunately I **Frankly** to them have
pian-jian.
prejudice

Supposing these hierarchies are indeed universal, the grammar should account for it. Moreover, in addition to strictly ordered adjuncts, ideally a model of adjunction should account for unordered adjuncts as well. For example, English PPs are unordered:

(7) a. The alliance officer shot Kaeli **in the cargo hold** *with a gun*.
b. The alliance officer shot Kaeli *with a gun* **in the cargo hold**.

It is not unusual to see this kind of asymmetry, where right adjuncts are unordered but left adjuncts are ordered.

4 Previous approaches to adjunction

This section provides a brief overview of four approaches to adjunction. The first two are from a categorial grammar perspective and account for the optionality and, more or less, transparency to selection; however, they are designed to model unordered adjuncts. The other two are MG formal-

[4]Data from Cinque (1999)

isations of the cartographic approach. Since the cartographic approach takes adjuncts to be regular selectors, unsurprisingly they account for order, but not easily for optionality or transparency to selection.

4.1 Categorial Grammar solutions

To account for the optionality and transparency, a common solution is for a modifier to combine with its modified phrase, and give the result the same category as the original phrase. In traditional categorial grammars, a nominal modifier has category N\N or N/N, meaning it combines with an N and the result is an N.

Similarly, in MGs, an X-modifier has features =XX: it selects an X and the resulting structure has category feature X.

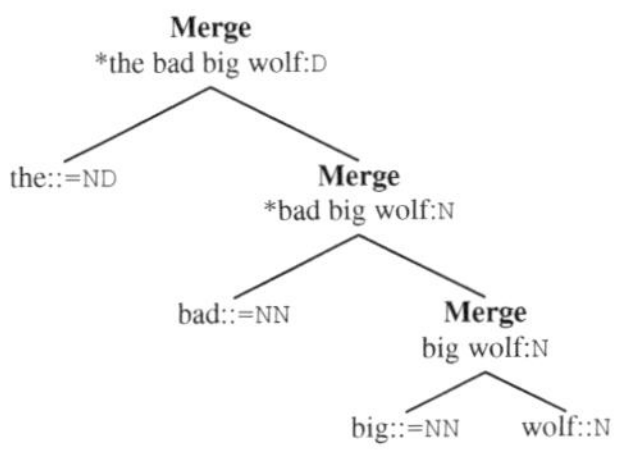

Figure 3: Traditional MG derivation of **the bad big wolf*

What this approach cannot account for is ordering. This is because the category of the new phrase is the same regardless of the modifier's place in the hierarchy. That is, the very thing that accounts for the optionality and the transparency of modifiers (that the category does not change) is what makes strict ordering impossible. Moreover, the modifier is not truly transparent to selection: the modifier in fact becomes the new head; it just happens to share a category with the original head. This can be seen in tree-generating grammars such as Stabler (1997) (Figure 4).

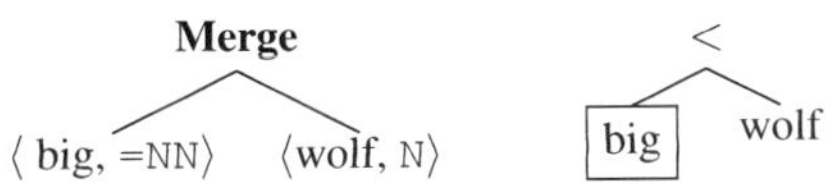

Figure 4: Derivation tree and derived *bare tree*. The $<$ points to the head, *big*.

4.1.1 Frey & Gärtner

Frey & Gärtner (2002) propose an improved version of the categorial grammar approach, one which keeps the modified element the head, giving true transparency to selection. They do this by asymmetric feature checking.

To the basic MG formalism a third polarity is added for **sel**, ≈X. This polarity drives the added function Adjoin. Adjoin behaves just like Merge except that instead of cancelling both ≈X and X, it cancels only ≈X, leaving the original X intact. This allows the phrase to be selected or adjoined to again by anything that selects or adjoins to X. This model accounts for optionality and true transparency: the modified element remains the head (Figure 4.1.1).

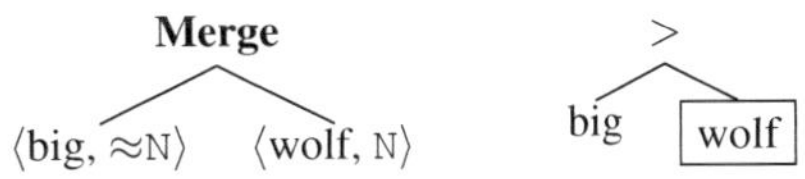

Figure 5: Frey & Gärtner: derivation tree and derived *bare tree*. The $>$ points to the head, *wolf*.

Since this grammar is designed to model unordered modifiers, illicit orders are also derivable (Figure 6).

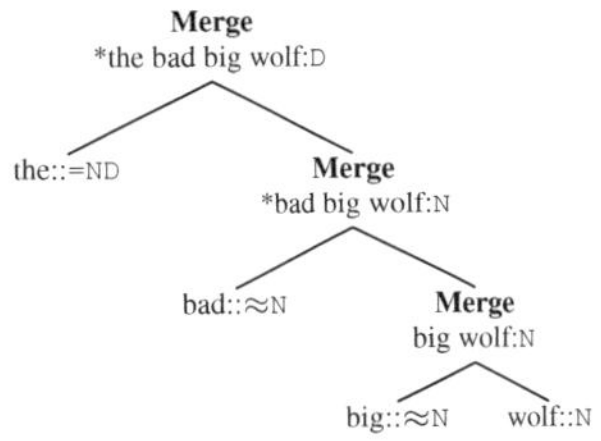

Figure 6: F & G derivation of **the bad big wolf*

4.2 Selectional approach

A third approach is to treat adjuncts just like any other selector. This is the approach taken by syntactic cartography. Such an approach accounts straightforwardly for order, but not for optionality or transparency; this is unsurprising since the phenomena I am modelling share only ordering restrictions with ordinary selection.

The idea is to take the full hierarchy of modifiers and functional heads, and have each select the one below it; for example, *big* selects *bad* but not vice versa, and *bad* selects *wolf*. However, here we are left with the question of what to do when *bad* is not present, and the phrase is just *the big wolf*. *big* does not select *wolf*.

4.2.1 Silent, meaningless heads

The first solution is to give each modifier and functional head a silent, meaningless version that serves only to tie the higher modifier to the lower.

For example, we add to the lexicon a silent, meaningless "size" modifier that goes where *big* and *small* and other LIs of category S go.

- ⟨ the, =S D⟩ ⟨ ϵ, =S D⟩
- ⟨ big, =G S⟩ ⟨ ϵ, =G S⟩
- ⟨ bad, =N G⟩ ⟨ ϵ, =N G⟩
- ⟨ wolf, N⟩

This solution doubles substantial portions of the lexicon. Doubling is not computationally significant, but it does indicate a missing generalisation: somehow, it just happens that each of these modifiers has a silent, meaningless doppelganger. Relatedly, the ordering facts are epiphenomenal. There is nothing forcing, say, D's to always select S's. There is no universal principle predicting the fairly robust cross-linguistic regularity.

Moreover, normally when something silent is in the derivation, we want to say it is contributing something semantically. Here these morphemes are nothing more than a trick to hold the syntax together. Surely we can do better.

4.2.2 Massive homophony

A second solution is for each morpheme in the hierarchy to have versions that select each level below it. For example, *the* has a version which selects N directly, one that selects "goodness" adjectives like *bad*, one that selects "size" adjectives like *big*, and indeed one for each of the ten or so levels of adjectives.

- ⟨the, =SD⟩ ⟨the, =GD⟩ ⟨the, =SD⟩ ⟨the, =ND⟩
- ⟨big, =GS⟩ ⟨big, =NatS⟩⟨big, =NS⟩
- ⟨bad, =NatG⟩ ⟨bad, =NG⟩
- ⟨Canadian, =NNat⟩
- ⟨wolf, N⟩

This second solution lacks the strangeness of silent, meaningless elements, but computationally it is far worse. To compute this we simply use Gauss's formula for adding sequences of numbers, since an LI at level i in a hierarchy has i versions. For example, in the model above, *the* is at level 4 (counting from 0), and there are 4 versions of *the*. For a lexicon Lex without these duplicated heads, and a language with k hierarchies of depths l_i for each $1 \leq i \leq k$, adding the duplicated heads increases the size of the lexicon. The increase is bounded below by a polynomial function of the depths of the hierarchies as follows:[5]

$$|\text{Lex}'| \geq \sum_{i=1}^{k} 1/2(l_i^2 + l_i) + |\text{Lex}|$$

5 Proposal

I propose a solution with three components: sets of categories defined to be adjuncts of particular categories, a partial order on **sel**, and a new operation Adjoin. The sets of adjuncts I base on Stabler (2013). The partial order models the hierarchies of interest (e.g. the Cinque hierarchy); Adjoin is designed to be sensitive to the order.

Adjoin operates on pairs of selectional features. The first element is the category of the first thing that was adjoined to, for example N. The second element is the category of the most recently adjoined element, for example $\mathtt{Adj}_3$. Adjoin is only defined if the new adjunct is higher in the hierarchy than the last adjunct adjoined.

I call these grammars Minimalist Grammars with Adjunction (MGAs).

Definition 5.1. A ***Minimalist Grammar with Adjunction*** is a six-tuple $G = \langle \Sigma, \langle \textbf{sel}, \geq \rangle, \textbf{ad}, \textbf{lic}, Lex, M \rangle$. Σ is a finite set called the *alphabet*. **sel**∪**lic** are finite sets of *base features*, and $\langle \textbf{sel}, \geq \rangle$ is a partial order. Let F={+f, -f, =X, [X, Y] | f∈ **lic**, X, Y ∈ **sel**}. **ad** : **sel** $\rightarrow \mathcal{P}(\textbf{sel})$ maps categories to their adjuncts. $Lex \subseteq \Sigma \cup \{\epsilon\} \times F^*$, and M is the set of operations **Merge, Move,** and **Adjoin**. The language L_G is the closure of Lex under M. A set $C \subseteq \textbf{sel}$ of designated features can be added; $\{[\mathsf{c},\mathsf{x}] | \mathsf{c} \in C, \mathsf{x} \in \textbf{sel}, x \geq c\}$ are the types of complete sentences.[6]

The differences between MGs defined above and MGAs are: (1) in MGAs **sel** is partially ordered; (2) in MGs the negative polarity for X ∈ **sel** is just X; in MGAs it is the pair [X,X]; (3) MGAs add a function: Adjoin; (4) MGAs define some subsets of **sel** to be adjuncts of certain categories; (5) Merge is redefined for the new feature pair polarity. (Move remains unchanged.)

[5] I say "bounded below" because this formula calculates the increase to the lexicon assuming there is exactly one LI at each level in the hierarchy. If there are more, each LI at level i of a hierarchy has i versions as well.

[6] I have replaced all negative selectional features X with pairs [X,X]. This is for ease of defining **Adjoin** and the new Merge. Equivalently, LIs can start with category features X as in a traditional MG, and Adjoin can build pairs. I chose the formulation here because it halves the number of cases for both Merge and Adjoin.

For $\langle A, \geq \rangle$ a partial order, $a, b \in A$ are ***incomparable***, written $a || b$, iff $a \not\geq b$ and $b \not\geq a$.

To shorten the definition of Adjoin, I define a function f_{adj} which determines the output features under Adjoin. If the adjunct belongs to the hierarchy of adjuncts being tracked by the second element of the feature pair, that second element changes. If not, the feature pair is unchanged.

Definition 5.2. For $\mathtt{W}, \mathtt{X}, \mathtt{Y}, \mathtt{Z} \in \textbf{sel}$, $\mathtt{W} \in \textbf{ad}(\mathtt{Y})$:

$$f_{\text{adj}}([\mathtt{W},\mathtt{X}],[\mathtt{Y},\mathtt{Z}]) = \begin{cases} [\mathtt{Y},\mathtt{W}] & \text{if } \mathtt{W} \geq \mathtt{Z} \\ [\mathtt{Y},\mathtt{Z}] & \text{if } \mathtt{W}||\mathtt{Z} \\ \text{undefined} & \text{otherwise} \end{cases}$$

Notice that if Z and W are incomparable, no record is kept of the feature (W) of the adjunct. This is just like Frey & Gärtner's asymmetric feature checking, and derives adjuncts that are unordered with respect to each other. In Definition 5.3, I model languages like English in which generally unordered adjuncts, like PPs, appear to the right, while ordered adjuncts, like adjectives, appear to the left. The rules could be easily modified for different orderings. See Section 6 for further discussion.

Definition 5.3 (*Adjoin*)**.** For s, t derived structures, $\gamma, \beta \in \{-\mathtt{f}|\mathtt{f} \in \textbf{lic}\}^*$, $\alpha \in \{+\mathtt{f}, =\mathtt{X}|\mathtt{f} \in \textbf{lic}, \mathtt{X} \in \textbf{sel}\}^*$, $\mathtt{W},\mathtt{X},\mathtt{Y},\mathtt{Z} \in \textbf{sel}$, $\mathtt{W} \in \textbf{ad}(\mathtt{Y})$, $C = f_{\text{adj}}([\mathtt{W},\mathtt{X}],[\mathtt{Y},\mathtt{Z}])$:

$$\textbf{Adjoin}(\langle s, [\mathtt{W},\mathtt{X}]\alpha\gamma\rangle{::}\text{mvrs}_s, \langle t, [\mathtt{Y},\mathtt{Z}]\beta\rangle :: \text{mvrs}_t) =$$

$$\begin{cases} \langle \textbf{Com}(s,t), \alpha C\rangle :: \text{mvrs}_s \cdot \text{mvrs}_t & \text{if } \gamma,\beta = \epsilon \;\&\; \mathtt{W} \geq \mathtt{Z} \\ \langle \textbf{Com}(t,s), \alpha C\rangle :: \text{mvrs}_s \cdot \text{mvrs}_t & \text{if } \gamma,\beta = \epsilon \;\&\; \mathtt{W}||\mathtt{Z} \\ \langle s, \alpha C\rangle :: \langle t, \beta\rangle :: \text{mvrs}_s \cdot \text{mvrs}_t & \text{if } \gamma = \epsilon, \beta \neq \epsilon \;\&\; \mathtt{W} \not< \mathtt{Z} \\ \langle t, \alpha C\rangle :: \langle s, \gamma\rangle :: \text{mvrs}_s \cdot \text{mvrs}_t & \text{if } \gamma \neq \epsilon, \beta = \epsilon \;\&\; \mathtt{W} \not< \mathtt{Z} \\ \langle \epsilon, \alpha C\rangle :: \langle s, \gamma\rangle :: \langle t, \beta\rangle :: \text{mvrs}_s \cdot \text{mvrs}_t & \text{if } \gamma,\beta \neq \epsilon \;\&\; \mathtt{W} \not< \mathtt{Z} \end{cases}$$

The first case is for ordered adjuncts where neither the adjunct nor the adjoined-to phrase will move (encoded in empty γ, β). The second is the same but for unordered adjuncts, which will appear on the right. The last three cases are for moving adjunct, moving adjoined-to phrase, and both moving, respectively. α is a sequence of positive licensing features, which allows adjuncts to take specifiers.

Merge needs a slight modification, to incorporate the paired categories. Notice that Merge is interested only in the first element of the pair, the "real" category.

Definition 5.4 (*Merge*)**.** For $\alpha, \beta \in F^*$, s, t derived structures, $\mathtt{X},\mathtt{Y} \in \textbf{sel}$:

$$\textbf{Merge}(\langle s, {=}\mathtt{X}\alpha\rangle :: \text{mvrs}_s, \langle t, [\mathtt{X},\mathtt{Y}]\beta\rangle{::}\text{mvrs}_t) = \begin{cases} (\textbf{Com}(s,t), \alpha) :: \text{mvrs}_s \cdot \text{mvrs}_t & \text{if } \beta = \epsilon \\ (s,\alpha) :: (t,\beta) :: \text{mvrs}_s \cdot \text{mvrs}_t & \text{if } \beta \neq \epsilon \end{cases}$$

Move remains as in definition 2.3 above.

5.1 Examples

MGAs are most easily understood by example. This first example demonstrates straightforward applications of Adjoin that derive strictly-ordered prenominal adjectives. *The big bad wolf* is derivable because the derivation remembers that an N-adjunct at level G in the hierarchy, ⟨bad, [G,G]⟩, adjoined to the noun. It encodes this fact in the second element of the pair [N,G]. *Big* is then able to adjoin because it too is an N-adjunct and it is higher in the hierarchy than *bad* (S>G). Finally, *the* can be defined to select *wolf* directly.

Let **sel** = $\{\mathtt{D},\mathtt{G},\mathtt{M},\mathtt{N},\mathtt{P},\mathtt{C},\mathtt{T},\mathtt{V}\}$ and the partial order $\geq$ on **sel** be such that $\mathtt{D} \geq \mathtt{S} \geq \mathtt{G} \geq \mathtt{M} \geq \mathtt{N}$ and $\mathtt{C} \geq \mathtt{T} \geq \mathtt{V}$

adjuncts = $\{\langle \mathtt{N}, \{\mathtt{S},\mathtt{G},\mathtt{M},\mathtt{P},\mathtt{C}\}\rangle\}$

Lex = {⟨bad, [G,G]⟩, ⟨big, [S,S]⟩, ⟨the, =N[D,D]⟩, ⟨wolf, [N,N]⟩, ⟨woods, [N,N]⟩, ⟨in, =D[P,P]⟩}

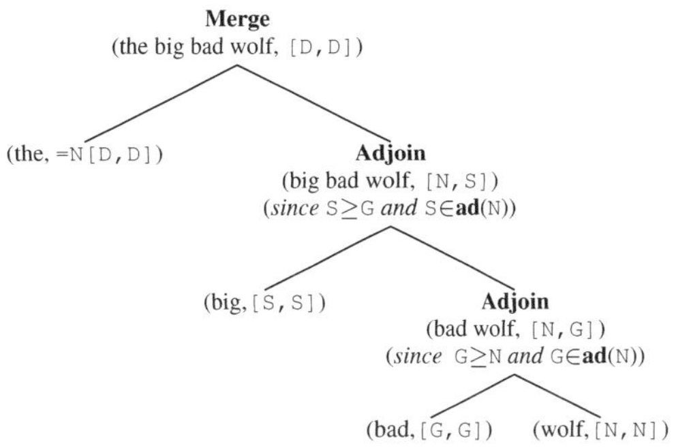

Figure 7: Valid derivation of *the big bad wolf*

**Bad big wolf*, on the other hand, is not derivable without movement since the derivation remembers that *big*, which is at level S in the hierarchy, has already been adjoined. *bad*, being lower in the hierarchy, cannot adjoin.

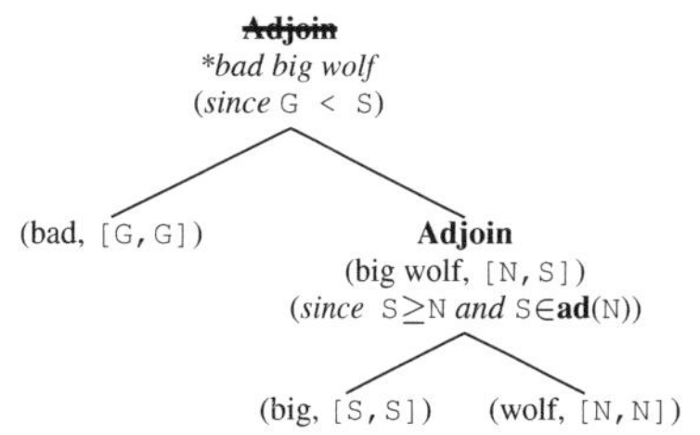

Figure 8: Invalid derivation of **bad big wolf*

This next example shows a right adjunct, a PP, being adjoined to an NP. Since P||N – that is, no hierarchical order is defined between N and P – the PP adjoins to the right, but does not alter the category of the noun.

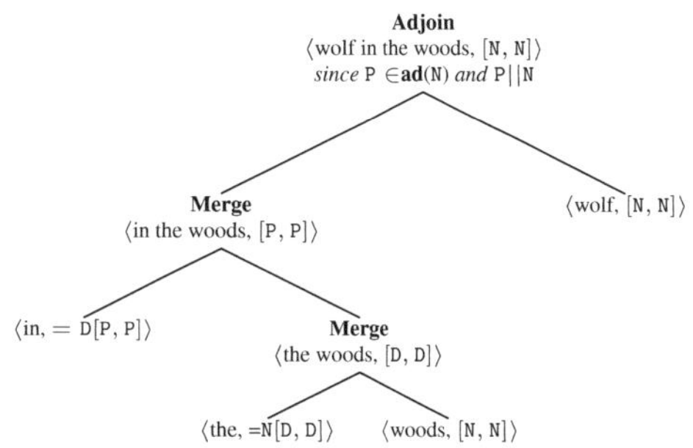

Figure 9: Right adjunction

6 Discussion and extensions

This model captures both the strict ordering of the merge-only models and the optionality and transparency to selection of the categorial approaches. Cinque's observation that there is a hierarchy of functional heads and adverbs is modelled directly by defining a hierarchy in the grammar itself. The strict linear order falls out of the order imposed on the selectional features and the definition of Adjoin: adjunction is only defined when the hierarchy is respected. Optionality is the result of the transitivity of orders: intervening adjuncts are not necessary for a higher one to be adjoined. Transparency to selection is modelled by the pairing of the selectional features: the original category of the modified element is preserved, and **Merge** can see only that feature. The adjuncts are literally ignored.

The cross-linguistic consistency of the orders is accounted for by the claim that all human languages have the same partial order on **sel**. As such, it does not have to be learned, but rather comes with the grammar.

Computationally, this approach has an advantage over the merge-only model with homophony as the latter increases the size of the lexicon by a polynomial function in the depths of the hierarchies of adjuncts, but the former does not.

6.1 Left and right adjuncts

As mentioned, I defined Adjoin to derive the asymmetry observed between left and right adjuncts in many languages: left adjuncts such as adverbs and descriptive adjectives are strictly ordered, while right adjuncts such as PPs and clauses are not. This fact is derived by letting the presence or absence of an ordering relation between the adjunct and modified category determine which case of Adjoin applies. If there is an order, the usual linear order will be calculated by **Com**, and the place in the hierarchy is tracked. Otherwise, the linear order is switched, and there is asymmetric feature checking.

If this is not the effect desired, there are alternatives. The simplest is to make the domain of the function **ad sel** $\times$ {right, left}, specifying the sets of right and left adjuncts. This allows for much more flexibility, for good or ill. It does not derive the asymmetry, but does allow ordered and unordered adjuncts to appear on the same side of the head, if such a pattern is desired. This is an empirical question.

6.2 Selection and adjuncts

This model allows LIs that are in the set of adjuncts to be selected normally as arguments, since adjuncts have categories of their own. For example, *Red Ridinghood was small* is derivable by allowing *was* to select ⟨small, [S,S]⟩: ⟨was, =S[V,V]⟩. This is an improvement over models that do not give adjuncts categories of their own, such as Frey & Gärtner's, but it is still lacking. In this model, there will have to be massive duplication in the lexicon so that *was* can select every adjective: ⟨was, =S[V,V]⟩, ⟨was, =G[V,V]⟩etc.

To solve this problem, we can take advantage of the function **ad**, and define *was* to select anything from a particular image under **ad**. Such a model expands the definition of Merge to operate not only on categories, but also on sets of categories. The model would look something like this:

Merge(⟨was, =**ad**(N)[V,V]⟩, ⟨small, [S,S]⟩) is defined iff S∈ **ad**(N)

Because the set of features F is finite, allowing Merge to be defined over subsets of F does not change the finite properties of MGs. Merge could in fact be allowed to be defined over any subset of F. I suggest this model because it is restricted: only sets that exist for other reasons already can be quantified over.

MGAs also allow adjuncts to select arguments and license Move. For example, a preposition can select a complement before becoming an adjunct PP. Moreover, a functional projection such as Focus can Move a focused phrase into its specifier from the main tree, or Topic can Merge a specifier. The latter is a result of allowing positive polarity features to follow the category pair. Recall that in traditional MGs, an LI must be of the following form for the derivation to succeed, where each p_i is a positive polarity feature, $\mathtt{X}, \mathtt{Y} \in \mathbf{sel}$ and each $\mathtt{f}_i \in \mathbf{lic}$:

$$(= \mathtt{Y}(p_1 p_2 ... p_n))\mathtt{X}(-\mathtt{f}_1 - \mathtt{f}_2 ... -\mathtt{f}_m)$$

However, in MGAs, LIs of the following form are possible if the LI will Adjoin, the crucial difference being the presence of $p_{n+1}...p_k$:

$$(= \mathtt{Y}(p_1 p_2 ... p_n))[\mathtt{X}, \mathtt{Y}](p_{n+1}...p_k)(-\mathtt{f}_1 - \mathtt{f}_2 ... -\mathtt{f}_m)$$

Figure 10 shows the end of a derivation in which the mover *briefly* is an adjunct, and so the licensor, the null Foc head. Its positive licensing feature +foc moves to the front of the stack of the derived structure's features.

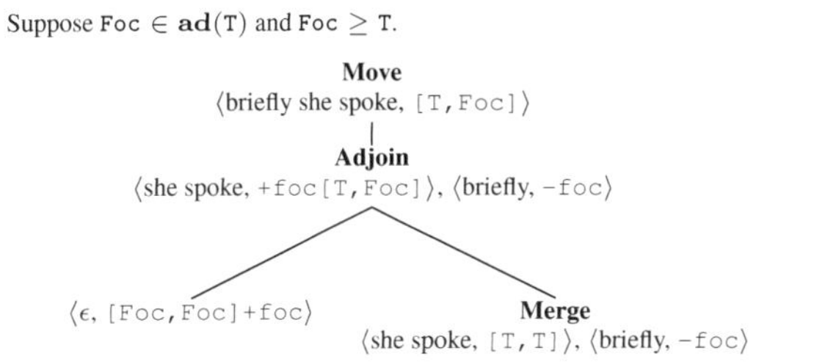

Figure 10: Adjunct FocP with moved specifier.

6.3 Adjuncts of adjuncts

In natural language, adjuncts can also be adjoined to, for example as in *the **very** bad wolf*. The function **ad** maps single categories to their adjuncts, but it is not generally the case that, say, an adverb, can only adjoin to certain adjectives. In order to capture this fact without duplication in the lexicon, Adjoin, like Merge, can be extended to allow subsets of F. Similarly to the Merge case, we can restrict these subsets by requiring that they be the image of a category under **ad**. For example:

⟨frankly, [Fr, Fr]⟩, ⟨unfortunately, [Fo, Fo]⟩, ⟨allegedly, [Al, Al]⟩, ⟨bad, [G, G]⟩, ⟨wolf, [N, N]⟩∈ *Lex*
Fr ≥ Fo ≥ Al ≥ V, S ≥ G ≥ N, P
ad (N) = {S, G, P}
ad (V) = **ad** (S) = **ad** (G) = {Fr, Fo, Al}

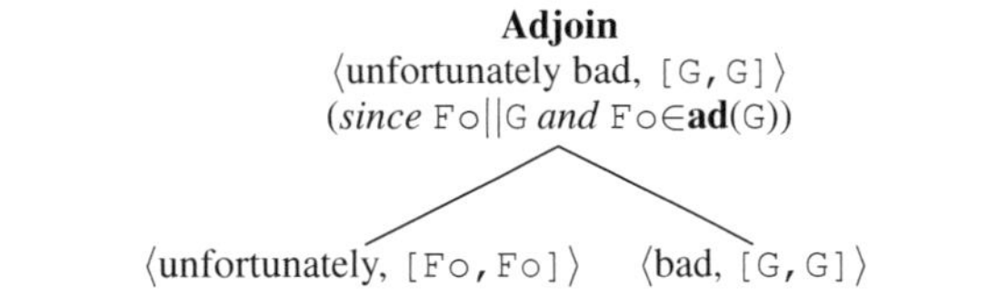

Figure 11: Adjoining to an adjunct

Notice however that we are still missing a generalisation: S, G, and indeed all adjectives have the same adjuncts. Now, this can be modelled by calling this set **ad**(**ad**(N)). However, such a solution assumes a special status for N over many other categories such as G: why **ad**(**ad**(N)) rather than **ad**(**ad**(G))? I would argue that such a status would reflect the reality of natural language. We can see N and V behaving in special ways: both are at the bottom of hierarchies, for example. However, as far as I am aware, no such status exists in any MGs. Formalising these observations is a matter for further research.

6.4 Islandhood

Adjuncts have another classic property: islandhood. Movement is not possible out of certain types of adjuncts.

(8) a. You left [because your ex showed up]$_{\text{Adj}}$
b. *Who did you leave [because ___ showed up]$_{\text{Adj}}$?

Any approach that keeps Adjoin separate from Merge introduces the option of stipulating the Adjunct Island Constraint (AIC), either as a separate constraint on Adjoin, as Frey & Gärtner do, or by simply not including movers$_s$ in the definition of Adjoin, making the function undefined when the adjunct carries movers. This is not very satisfying, though: better perhaps would be to derive it, as Graf (2013) does. On the other hand, perhaps not all adjuncts are islands. If *beside* ___ is an adjunct in (9), it is not an adjunct island.

(9) Who are you sitting [beside ___]$_{\text{Adjunct}}$?

As always, islands must remain a matter for further research.

7 Conclusion

I have presented a model of adjunction that accounts for both the optionality and the strict ordering of many adjuncts. MGAs accomplish this by the simple expedience of keeping track of two pieces of information at once: the original category of the projecting phrase, and the category of the most recent adjunct to adjoin. This allows Adjoin to be defined to only apply when the next adjunct is not lower in a hierarchy than the last. At the same time, Merge can see the original category, and ignores the adjunct's category.

I have also suggested some extensions of MGAs to more efficiently account for adjuncts as the second argument of Merge and Adjoin. These involved quantification over categories, with the added suggestion that the sets of categories in question be restricted by the sets of adjuncts already defined.

Future directions for this research include not only matters internal to the model, such as how best to model adjuncts of adjuncts, but also larger questions of the mathematical properties of MGAs. MGAs are weakly equivalent to MGs, since MGAs merely take existing ways to derive certain strings and seek more efficient ways, which capture more generalisations. If every adjunct in the lexicon is replaced with the right set of selectors, Adjoin does not need to be used. For example, the adjectives in the MGA lexicon used in the examples in Section 5.1 can be replaced by the adjectives in either grammar from the selectional approaches in Section 4.2, and the same string set can be generated.

Clearly MGs and MGAs are not strongly equivalent: the derivation trees differ in that MGAs have a function that is not present in MGs.

Because the possible configurations of features remains finite, the derivation tree languages of MGAs should prove to be regular, following Kobele et al (2007)'s presentation: transition rules for Adjoin need merely be added.

Also of interest are the subregular properties of the derivation tree language. Although to my knowledge such notions as *tierwise strictly local* (Heinz et al., 2011) have not yet been formally defined for tree languages, I conjecture that in MGAs, Merge is tierwise strictly k-local, and Adjoin is strictly k-local.

References

Noam Chomsky. 1995. *The Minimalist Program*. MIT Press, Cambridge, MA.

Gugliemo Cinque. 1999. *Adverbs and functional heads: a cross-linguistic perspective*. Oxford studies in comparative syntax. Oxford University Press, Oxford.

Gugliemo Cinque. 2010. *The syntax of adjectives: a comparative study*. Linguistic Inquiry monographs. MIT Press, Cambridge, MA.

Meaghan Fowlie. 2011. Multidominant minimalist grammars. Master's thesis, University of California, Los Angeles.

Werner Frey and Hans-Martin Gärtner. 2002. On the treatment of scrambling and adjunction in minimalist grammars. In *Proceedings of the Conference on Formal Grammar (FGTrento)*, pages 41–52, Trento.

Thomas Graf. 2013. The price of freedom: Why adjuncts are islands. Slides of a talk given at the Deutsche Gesellschaft für Sprachwissenschaft 2013, March 12–15, University of Potsdam, Potsdam, Germany.

Jeffrey Heinz, Chetan Rawal, and Herbert Tanner. 2011. Tier-based strictly local constraints for phonology. In *Proceedings of the 49th Annual Meeting of the Association for Computational Linguistics*, Portland, Oregon, USA, June. Association for Computational Linguistics.

Edward L. Keenan and Edward P. Stabler. 2003. *Bare Grammar*. CSLI Publications, Stanford.

Gregory M. Kobele, Christian Retoré, and Sylvain Salvati. 2007. An automata-theoretic approach to minimalism. In J. Rogers and S. Kepser, editors, *Model Theoretic Syntax at ESSLLI '07*. ESSLLI.

Luigi Rizzi. 2004. Locality and left periphery. In Adriana Belletti, editor, *Structures and Beyond: The Cartography of Syntactic Structures*, volume 3, pages 223–251. Oxford University Press, Oxford.

Edward Stabler. 1997. Derivational minimalism. *Logical Aspects of Computational Linguistics*, pages 68–95.

Edward Stabler. 2013. Bracketing paradoxes and opacity: Parsing late adjunction in copy constructions. Talk given at UCLA Mathematical Linguistics Circle, April.

On the Parameterized Complexity of Linear Context-Free Rewriting Systems

Martin Berglund
Umeå University, Sweden
mbe@cs.umu.se

Henrik Björklund
Umeå University, Sweden
henrikb@cs.umu.se

Frank Drewes
Umeå University, Sweden
drewes@cs.umu.se

Abstract

We study the complexity of uniform membership for Linear Context-Free Rewriting Systems, i.e., the problem where we are given a string w and a grammar G and are asked whether $w \in \mathcal{L}(G)$. In particular, we use parameterized complexity theory to investigate how the complexity depends on various parameters. While we focus primarily on rank and fan-out, derivation length is also considered.

1 Introduction

Linear Context-Free Rewriting Systems (LCFRS) were introduced by Vijay-Shanker et al. (1987) with the purpose of capturing the syntax of natural language.[1] It is one of several suggested ways of capturing Joshi's concept of *mildly context-sensitive languages* (Joshi, 1985). As such, it strengthens the expressive power of context-free grammars, while avoiding the full computational complexity of context-sensitive grammars.

One of the defining features of mildly context-sensitive languages is that they should be decidable in polynomial time. This is indeed true for every language that can be generated by an LCFRS. Unlike the case for context-free grammars, however, the *universal* or *uniform* membership problem for LCFRSs, where both the grammar and the string in question are considered as input, is known to be PSPACE-complete (Kaji et al., 1992), making a polynomial time solution very improbable.

The best known algorithms for the problem have a running time of $\mathcal{O}(|G| \cdot |w|^{f \cdot (r+1)})$, where G is the grammar, w is the string, f is the *fan-out* and r is the *rank* of the grammar (Seki et al., 1991; Burden and Ljunglöf, 2005; Boullier, 2004). (For a definition of fan-out and rank, see Section 2.) Unlike the rank of a context-free grammar, the fan-out and rank of an LCFRS cannot in general be reduced to some fixed constant. Increasing the fan-out always gives more generative power, as does increasing the rank while keeping the fan-out fixed (Satta, 1998). The rank can be reduced to two, but at the price of an exponential increase in the fan-out.

Research into algorithms for LCFRS parsing that are efficient enough for practical use is quite active. For example, algorithms for restricted cases are being studied, e.g., by Gómez-Rodríguez et al. (2010), as well as rank reduction, primarily in special cases, where the fan-out is not affected; see, e.g., Sagot and Satta (2010).

This article is a first step towards a finer computational complexity analysis of the membership problem for LCFRSs. Specifically it asks the question "could there exist an algorithm for the uniform LCFRS membership problem whose running time is a fixed polynomial in $|w|$ times an arbitrary function in f and r?" By employing parameterized complexity theory, we show that such an algorithm is very unlikely to be found. Fixing the rank of the grammar to one, the membership problem, parameterized by the fan-out, is W[SAT]-hard. Fixing the fan-out to two and taking the rank as the parameter, the problem is W[1]-hard. Finally, if the fan-out, rank, *and derivation length* are included in the parameter, the problem is W[1]-complete. These results help guide future work, suggesting other types of parameters and grammar restrictions that may yield more favorable complexity results.

2 Preliminaries

For $n \in \mathbb{N}$, we write $[n]$ for $\{1, \ldots, n\}$ and $[n]_0$ for $\{0\} \cup [n]$. Given an alphabet Σ we write Σ^* for all strings over Σ and Σ^+ for all non-empty strings. The empty string is denoted by ε.

[1] Seki et al. (1991) independently suggested the nearly identical Multiple Context-Free Grammars.

2.1 Linear context-free rewriting systems

Let Σ be an alphabet, $x_1, \ldots, x_n$ variables, and $w_1, \ldots, w_k$ strings over Σ such that

$$w_1 \cdots w_k = \alpha_0 \cdot x_{\pi(1)} \cdot \alpha_1 \cdots x_{\pi(n)} \cdot \alpha_n$$

for some permutation π and some strings $\alpha_0, \ldots, \alpha_n \in \Sigma^*$. Then define f as a function over tuples of strings such that

$$f((x_1, \ldots), \ldots, (\ldots, x_n)) = (w_1, \ldots, w_k).$$

A function is *linear regular* if and only if it can be described in this way. For example $f((x_1), (x_2)) = (a, bx_2x_1c)$ is linear regular, and $f((aaa), (bc)) = (a, bbcaaac)$.

Definition 2.1. A *Linear Context-Free Rewriting System* is a tuple $G = (N, \Sigma, F, R, S)$, where N is an alphabet of *nonterminals*, where each $A \in N$ has an associated *fan-out* $\#(A)$; $S \in N$ is the initial nonterminal with $\#(S) = 1$; Σ is an alphabet of *terminals*; F is a set of linear regular functions; and R is a set of rules of the form $A \rightarrow g(B_1, \ldots, B_n)$, where $A, B_1, \ldots, B_n \in N$ and g is a function in F of type

$$(\Sigma^*)^{\#(B_1)} \times \cdots \times (\Sigma^*)^{\#(B_n)} \rightarrow (\Sigma^*)^{\#(A)}.$$

For rules $A \rightarrow g()$, where g has arity 0 and $g() = (\alpha_1, \ldots, \alpha_{\#(A)})$, we often simply write $A \rightarrow (\alpha_1, \ldots, \alpha_{\#(A)})$.

The *rank* of a rule is the number of nonterminals on the right-hand side. The rank of G is the maximal rank of any rule in R. The *fan-out* of G is the maximal fan-out of any nonterminal in N.

The language generated by a nonterminal A is a set of n-tuples, where $n = \#(A)$.

Definition 2.2. Let $G = (N, \Sigma, F, R, S)$ be a linear context-free rewriting system. Let $\mathcal{L}_A \subseteq (\Sigma^*)^{\#(A)}$ denote the tuples that a nonterminal $A \in N$ can generate. This is the smallest set such that if $A \rightarrow f(B_1, \ldots, B_n)$ is in R then, for all $b_i \in \mathcal{L}_{B_i}$ where $i \in [n]$, $f(b_1, \ldots, b_n) \in \mathcal{L}_A$. The language of G is $\mathcal{L}(G) = \mathcal{L}_S$.

For $i \in \mathbb{N}$, we write i-LCFRS for the class of all LCFRSs of rank at most i and LCFRS(i) for the class of all LCFRSs of fan-out at most i. We also write i-LCFRS(j) for i-LCFRS$\cap$LCFRS(j).

2.2 Parameterized complexity theory

We only reproduce the most central definitions of parameterized complexity theory. For a more thorough treatment, we refer the reader to (Downey and Fellows, 1999; Flum and Grohe, 2006).

A *parameterized problem* is a language $\mathcal{L} \subseteq \Sigma^* \times \mathbb{N}$, where Σ is a finite alphabet. The second component is called the *parameter*. An algorithm for $\mathcal{L}$ is *fixed-parameter tractable* if there is a computable function f and a polynomial p such that for every $(x, k) \in \Sigma^* \times \mathbb{N}$, the algorithm decides in time $f(k) \cdot p(|x|)$ whether $(x, k) \in \mathcal{L}$. The problem of deciding $\mathcal{L}$ is fixed-parameter tractable if there is such an algorithm. If so, $\mathcal{L}$ belongs to the class FPT.

A parameterized problem $\mathcal{L} \subseteq \Sigma^* \times \mathbb{N}$ is *fpt-reducible* to another parameterized problem $\mathcal{K} \subseteq \Gamma^* \times \mathbb{N}$ if there is a mapping $R : \Sigma^* \times \mathbb{N} \rightarrow \Gamma^* \times \mathbb{N}$ such that

1. for all $(x, k) \in \Sigma^* \times \mathbb{N}$, we have $(x, k) \in \mathcal{L}$ if and only if $R(x, k) \in \mathcal{K}$,
2. there is a computable function f and a polynomial p such that $R(x, k)$ can be computed in time $f(k) \cdot p(|x|)$, and
3. there is a computable function g such that for every $(x, k) \in \Sigma^* \times \mathbb{N}$, if $R(x, k) = (y, k')$, then $k' \leq g(k)$.

Note that several parameters may be combined into one by taking their maximum (or sum).

The most commonly used hierarchy of parameterized complexity classes is the following.

$$\text{FPT} \subseteq \text{W[1]} \subseteq \text{W[2]} \subseteq \cdots \subseteq \subseteq \text{W[SAT]} \subseteq \text{W[P]} \subseteq \text{XP}$$

The classes W[1],...,W[P] are defined using circuits or, alternatively, logic. None of the inclusions is known to be strict, except that FPT is a strict subclass of XP. It is widely believed, however, that each of them is strict. The class XP is the class of all parameterized problems for which there is a computable function f such that every instance (x, k) can be decided in time $|x|^{f(k)}$.

2.3 Problems of interest

We know from Kaji et al. (1992) that the universal membership problem for 1-LCFRSs is PSPACE-complete. Satta (1992) has further shown that LCFRS(2)-MEMBERSHIP is NP-hard.

We study the following decision problems, where the symbol $\mathfrak{P}$ is used to indicate what the parameter is:

- $\mathfrak{P}$-LCFRS(j)-MEMBERSHIP, where $j \in \mathbb{N}$ is the membership problem for LCFRS(j), parameterized by the rank.

- i-LCFRS($\mathfrak{P}$)-MEMBERSHIP, where $i \in \mathbb{N}$ is the membership problem for i-LCFRS, parameterized by the fan-out.
- $\mathfrak{P}$-LCFRS($\mathfrak{P}$)-MEMBERSHIP is the membership problem for LCFRS parameterized by the rank and the fan-out.
- SHORT $\mathfrak{P}$-LCFRS($\mathfrak{P}$)-DERIVATION is the membership problem for LCFRS parameterized by the rank, the fan-out, and the derivation length.

Since there are algorithms that solve the membership problem for LCFRSs with rank r and fan-out t and string w in time $|w|^{(r+1)t}$ (see, e.g., (Seki et al., 1991; Burden and Ljunglöf, 2005; Boullier, 2004)), we can immediately conclude that $\mathfrak{P}$-LCFRS($\mathfrak{P}$)-MEMBERSHIP, as well as every other parameterized membership problem mentioned above, belongs to XP.

3 Fixed rank grammars

The following theorem establishes a lower bound for 1-LCFRSs parameterized by the fan-out.

Theorem 3.1. 1-LCFRS($\mathfrak{P}$)-MEMBERSHIP *is* W[SAT]*-hard.*

The proof of Theorem 3.1 is by reduction from WEIGHTED MONOTONE SATISFIABILITY. Before we get into the actual proof, we discuss some properties of this problem.

Definition 3.1. A *monotone Boolean formula* is a Boolean formula that contains only conjunctions, disjunctions, and variables. In particular, there are no negations. An instance of WEIGHTED MONOTONE SATISFIABILITY is a pair (ϕ, k), where ϕ is a monotone Boolean formula and $k \in \mathbb{N}$. The question is whether ϕ has a satisfying assignment of weight k, i.e., an assignment that sets exactly k of the variables that occur in ϕ to true. The parameter is k. WEIGHTED MONOTONE SATISFIABILITY is W[SAT]-complete (Abrahamson et al., 1993; Abrahamson et al., 1995; Downey and Fellows, 1999).

We can view a monotone Boolean formula ϕ as an unranked tree, where the root node corresponds to the top level clause and the leaves correspond to bottom level clauses, i.e., variable occurrences. The set $pos(\phi)$ of *positions* of ϕ is defined as usual, consisting of strings of natural numbers that indicate how to navigate to the clauses in a tree representation of ϕ. We denote each subclause of ϕ by C_s, where $s \in pos(\phi)$ is its position. Thus

$$\phi = (((x_1 \wedge (x_2 \vee x_3)) \vee x_3 \vee (x_3 \wedge x_4)) \wedge \\ \wedge x_2 \wedge ((x_1 \wedge (x_2 \vee x_4)) \vee (x_1 \wedge x_3)))$$

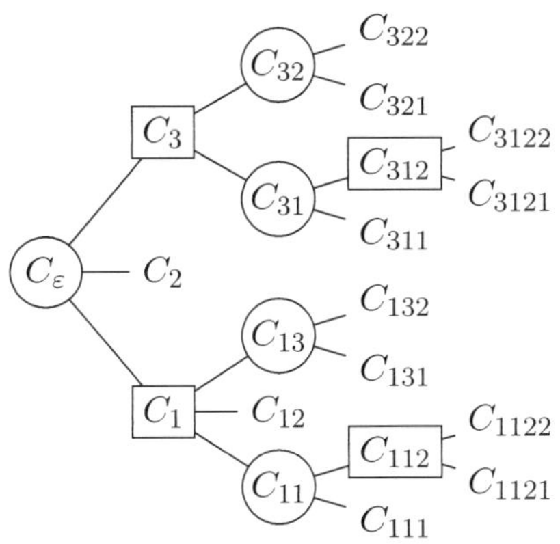

Figure 1: A formula ϕ and its tree representation. Conjunctive clauses are round and disjunctive rectangular. For example, C_{111} is the leftmost occurrence of x_1 and C_{13} the clause $(x_3 \wedge x_4)$.

C_ε denotes the whole of ϕ, while, e.g., C_{ijl} is the lth clause of the jth clause of the ith clause of ϕ. See Figure 1 for an example. We use $\mathcal{C}$ for the set of all clauses of ϕ and $\mathcal{C}_\wedge, \mathcal{C}_\vee$, and $\mathcal{C}_{Var}$ for the sets of conjunctive, disjunctive, and bottom level clauses, respectively. For all $c \in \mathcal{C}_{Var}$ let $Var(c)$ denote the variable in c, and let $Var(\phi)$ denote the set of all variables in ϕ.

Given a monotone Boolean formula ϕ and a variable assignment $\rho\colon Var(\phi) \rightarrow \mathbb{B}$, we define a *verification tour* for ϕ and ρ. Such a tour moves through the tree representation of ϕ, starting at the root node, and verifies that ρ satisfies ϕ. To this end, we first define the function $Next : pos(\phi) \rightarrow pos(\phi) \cup \{True\}$ as follows. For the root clause let $Next(\varepsilon) = True$. For all $si \in pos(\phi)$, where $s \in \mathbb{N}^*$ and $i \in \mathbb{N}$, if $C_s \in \mathcal{C}_\wedge$ and $s(i+1) \in pos(\phi)$ let $Next(si) = s(i+1)$, otherwise let $Next(si) = Next(s)$.

A verification tour over ϕ, given a variable assignment ρ is constructed by the following procedure. Set the initial position $p = \varepsilon$, then

- If $C_p \in \mathcal{C}_\wedge$ set $p \leftarrow p1$ (i.e., go to the first subclause).
- If $C_p \in \mathcal{C}_\vee$ set $p \leftarrow pi$ for any i (i.e. nondeterministically pick a subclause).
- If $C_p \in \mathcal{C}_{Var}$ verify that $\rho(Var(C_p)) = true$. If so, set $p \leftarrow Next(p)$ and repeat. Otherwise, the verification tour fails.

A verification tour succeeds if it reaches $True$.

The following lemma can be proved by straightforward induction on the structure of ϕ.

Lemma 3.2. *If a verification tour for ϕ and variable assignment ρ succeeds, then ρ satisfies ϕ.*

We are now ready to prove Theorem 3.1.

Proof of Theorem 3.1. Let (ϕ, k) be an instance of WEIGHTED MONOTONE SATISFIABILITY. Let $\{x_1, \ldots, x_n\}$ be the variables that appear in ϕ. In particular, n is the number of distinct variables. Let m be the number of bottom level clauses.

Intuitively, the LCFRS we will construct will guess a weight k variable assignment ρ and then simulate a verification tour for ϕ and ρ.

Basically, we will use one nonterminal per clause and use the structure of the grammar to simulate a verification tour. In order to verify that the necessary bottom level clauses can all be satisfied through the same k true variables, we will use the input string to be parsed. The string w will consist of bracketed sequences of m copies of each of the n variables, i.e., $w = [x_1^m] \cdots [x_n^m]$. To understand the construction of the grammar, please keep in mind that the only derivations that matter are those generating this particular input string.

The grammar will guess which k variables should be set to true and disregard the other variables. Technically, this is done by first letting a nonterminal F generate a tuple of $k+1$ strings $s_0, \ldots, s_k$ such that each s_i consists of zero or more of the bracketed sequences of variables to be disregarded. The rest of the grammar generates exactly k bracketed sequences that will be interleaved with $s_0, \ldots, s_k$. During the generation of these k bracketed sequences it is nondeterministically verified that the corresponding truth assignment satisfies ϕ.

We use the following set of nonterminals:

$$\{S, F\} \cup \{C_s \mid s \in pos(\phi) \cup \{True\}\}$$

For S, there is only one rule: $S \to f_S(F)$. The function f_S places brackets around the k variables that are guessed to be true, represented by the strings $t_1, \ldots, t_k$, and interleaves them with the remaining variables, represented by the strings $s_0, \ldots, s_k$:

$$f_S(s_0, \ldots, s_k, t_1, \ldots, t_k) = (s_0[t_1]s_1 \cdots [t_k]s_k)$$

The nonterminal F has rules $F \to f_{F,i,j}(F)$ for all $i \in [n]$ and $j \in [k]_0$. These rules produce the bracketed sequences of copies of the variables x_i to be disregarded, as can be seen from the corresponding function:

$$f_{F,i,j}(s_0, \ldots, s_k, t_1, \ldots, t_k) = (s_0, \ldots, s_j[x_i^m], \ldots, s_k, t_1, \ldots, t_k)$$

Moreover, there is a single rule

$$F \to f_F(C_\varepsilon)$$

with

$$f_F(t_1, \ldots, t_k) = (\varepsilon, \ldots, \varepsilon, t_1, \ldots, t_k)$$

The rules for the nonterminals that represent clauses differ according to the type of the clause, i.e., if the nonterminal represents a conjunctive clause, a disjunctive clause, or a variable. For each conjunctive clause C_s there is exactly one rule, representing a move to its first subclause. Here, f_{id} is the identity function.

$$C_s \to f_{id}(C_{s1})$$

For every disjunctive clause C_s and every i such that C_{si} is a subclause of C_s there is one rule.

$$C_s \to f_{id}(C_{si})$$

For every bottom level clause, i.e., $C_s \in \mathcal{C}_{Var}$, every $i \in [k]$ and every $j \in [m]$ there is one rule.

$$C_s \to f_{s,i,j}(C_{Next(s)})$$

Intuitively, such a rule corresponds to producing j copies of the variable of clause C_s in component i of the tuple and moving on to the next clause that should be visited in a verification tour. This can be seen from the corresponding function.

$$f_{s,i,j}(t_1, \ldots, t_k) = (t_1, \ldots, Var(C_s)^j t_i, \ldots, t_k)$$

The reason that the function produces j copies of the variable, rather than just one, is that it is unknown beforehand how many times a bottom level clause that represents that particular variable will be visited. Thus the number of copies to be produced has to be guessed nondeterministically in order to make sure that a total of m copies of each variable set to true are eventually produced.

If there is a weight k satisfying assignment, there will also be a verification tour that eventually reaches $True$ when $Next$ is called (by Lemma 3.2). The single rule for C_{True} simply produces a k-tuple of empty strings.

The reduction is polynomial and the fan-out of the resulting grammar is $2k+1$. Thus it is an FPT-reduction. It remains to argue that the grammar can produce w if and only if ϕ has a satisfying assignment of weight k.

We first note that whatever tuple is derived from F, the first $k+1$ entries in the tuple consist of bracketed sequences of the form $[x_l^m]$. If the grammar can produce w, it follows that the tuple $(t_1, \ldots, t_k)$ produced from C_ε must be such that each t_i equals m copies of the same variable name.

Any successful derivation of a string by the grammar corresponds to a verification tour of ϕ and the variable assignment that sets the variables that appear in $(t_1, \ldots, t_k)$ to true and all other variables to false. Thus ϕ has a satisfying assignment of weight k.

For the other direction, assume that ϕ has a satisfying assignment of weight k. Then the grammar can guess this assignment and a corresponding successful verification tour, thus producing w. □

Note that Theorem 3.1 can easily be strengthened to grammars with a binary terminal alphabet. It is enough to represent each variable name by a bitstring of length $\lceil \log_2(m) \rceil$ in the above reduction. We also note that Theorem 3.1 immediately implies that $\mathfrak{P}$-LCFRS($\mathfrak{P}$)-MEMBERSHIP is W[SAT]-hard.

4 Fixed fan-out grammars

We next turn to the case where the fan-out is fixed to two, while the rank is treated as a parameter.

Theorem 4.1. *$\mathfrak{P}$-LCFRS(2)-MEMBERSHIP is W[1]-hard.*

Proof. We reduce from k-CLIQUE, the problem of deciding whether a given graph has a clique of size k, with k as the parameter. This problem is known to be W[1]-complete (Flum and Grohe, 2006). Let $G = (\mathsf{V}, \mathsf{E})$ be an undirected graph. We assume, without loss of generality, that $\mathsf{V} = \{1, \ldots, n\}$ and that an edge connecting nodes $i, j \in \mathsf{V}$ is represented as the ordered pair (i, j) such that $i < j$, i.e., $\mathsf{E} \subseteq \{(i,j) \in \mathsf{V} \times \mathsf{V} \mid i < j\}$. To find out whether G has a clique of size k we construct an instance of the membership problem for LCFRSs.

The input alphabet is $\Sigma = \{0, 1\}$. Construct the input string as

$$w = \underbrace{0^n 10^n 10^n 1 \cdots 10^n}_{(3k+2)(k-1)/2 \text{ ones}}.$$

The nonterminals are $N = \{A, E, C, S\}$, with S being the initial nonterminal. The rules are the following.

$\{A \to 0^i \mid i \in \{1, \ldots, n\}\}$.

$\{E \to 0^{n-i}10^{n-j} \mid (i,j) \in \mathsf{E}\}$.

$\{C \to (0^i, 0^{n-i}10^i) \mid i \in \{1, \ldots, n\}\}$.

Handling S is a bit more complex. Let $\phi = k(k-1)/2$, the number of edges in a k-clique. Then the unique rule for S is:

$$S \to f(\underbrace{E, \ldots, E}_{\phi}, \underbrace{C, \ldots, C}_{2\phi}, \underbrace{A, \ldots, A}_{2k}).$$

Now we need to define f. Consider the following application of f.

$$f(e_1, \ldots, e_\phi, (c_1, \hat{c}_1), \ldots, (c_\phi, \hat{c}_\phi), \\ (d_1, \hat{d}_1), \ldots, (d_\phi, \hat{d}_\phi), a_1, \ldots, a_{2k}).$$

The application above evaluates to the string

$$c_1 e_1 d_1 1 c_2 e_2 d_2 1 \cdots \\ \cdots 1 c_\phi e_\phi d_\phi 1 a_1 \theta_1 a_2 1 a_3 \theta_2 a_4 1 s 1 a_{2k-1} \theta_k a_{2k}.$$

The substrings θ_1 through θ_k are left to be defined, and will contain all the $\hat{c}$ and $\hat{d}$ arguments in a careful configuration derived from the structure of a clique. Let $(\pi_1, \pi'_1), \ldots, (\pi_\phi, \pi'_\phi)$ be the lexicographically sorted sequence of edges in a k-clique with nodes numbered 1 through k. For example, $(\pi_1, \pi'_1) = (1,2)$, $(\pi_2, \pi'_2) = (1,3)$, $(\pi_k, \pi'_k) = (2,3)$, and $(\pi_\phi, \pi'_\phi) = (k-1, k)$. Then, for each l, find the longest subsequences $i_1, \ldots, i_p$ and $j_1, \ldots, j_q$ of $1, \ldots, \phi$ for which $\pi_{i_1} = \cdots = \pi_{i_p} = l$ and $\pi'_{j_1} = \cdots = \pi_{j_q} = l$, and let $\theta_l = \hat{c}_{i_1} \cdots \hat{c}_{i_p} \hat{d}_{j_1} \cdots \hat{d}_{j_q}$. □

This construction is simpler than it may at first appear. Basically, the clique is found by generating $k(k-1)/2$ copies of E, each of which will be placed so that it has no choice but to generate an edge in a k-clique. Looking at the first part of the string, each $1c_l e_l d_l 1$ must generate a string of the form $10^n 10^n 1$: e_l will generate some $0^{n-i}10^{n-j}$, were (i,j) is an edge in G, which forces c_l to generate 0^i and d_l to generate 0^j. The trick is that c_l and d_l yield the first string in a *pair* generated by an instance of C. The *other* string in the pair describes the same number as the first, but in such a way that it can be carefully placed in the latter part of the derivation string, thus forcing other instances of the C nonterminal to pick *the same* node (number of zeros) to generate. These are then placed in such a way that the edges picked by the instances of E all belong to the same clique. For example, for $k = 3$ the result of f will be $c_1e_1d_11c_2e_2d_21c_3e_3d_31a_1c_1c_2a_21c_3d_11d_2d_3$, where the latter part ensures that c_1 and c_2 have to pick the same node (lowest-numbered node in the clique), as do c_3 and d_1, and d_2 and d_3.

5 Short derivations

In this section, we consider the length of derivations as an additional parameter. As usual, the length of a derivation is the number of derivation steps it consists of. (In a derivation of an LCFRS (N, Σ, F, R, S), this is the same as the number of applications of functions in F.)

Let $G = (N, \Sigma, F, R, S)$ be an LCFRS in the following. Consider the following problem:

Definition 5.1. An instance of the SHORT $\mathfrak{P}$-LCFRS($\mathfrak{P}$) DERIVATION problem consists of a LCFRS G, some $w \in \Sigma^*$ and a constant $d \in \mathbb{N}$. The question asked is: can w be derived by G in at most d steps? The parameter is $k = d + r + f$ where r is the maximum rank and f the maximum fanout.

Lemma 5.1. SHORT $\mathfrak{P}$-LCFRS($\mathfrak{P}$) DERIVATION *is* W[1]*-hard.*

Proof. The W[1]-hardness of the problem follows immediately from the reduction in the proof of Theorem 4.1, since k-Clique is reduced to an instance of LCFRS membership with $\mathcal{O}(k^2)$ derivation steps, rank $\mathcal{O}(k^2)$, and fixed fan-out. □

We next demonstrate that SHORT $\mathfrak{P}$-LCFRS($\mathfrak{P}$) DERIVATION is *in* W[1] (and is therefore W[1]-complete) by reducing to SHORT CONTEXT-SENSITIVE DERIVATION, shown to be W[1]-complete by Downey et al. (1994). Let $H = (N_H, \Sigma_H, R_H, S_H)$ be an arbitrary context-sensitive grammar in the following. A context-sensitive grammar has nonterminals, terminals and a starting nonterminal just like a LCFRS, but the rules are of the form $\alpha \rightarrow \beta$ for strings $\alpha, \beta \in (\Sigma_H \cup N_H)^*$ where $0 < |\alpha| \leq |\beta|$. A derivation starts with the string S_H. A string $w \cdot \alpha \cdot w'$ can be turned into $w \cdot \beta \cdot w'$ in one derivation step if $(\alpha, \beta) \in R_H$.

Definition 5.2. An instance of the SHORT CONTEXT-SENSITIVE DERIVATION problem consists of a context-sensitive grammar H, a string $w \in \Sigma_H^*$, and a constant $d_H \in \mathbb{N}$. The question is: can w be derived by H in at most d_H steps? The parameter is d_H.

We are now ready to prove membership in W[1] by a FPT-reduction from (G, w, d) to (H, w, d_H).

Lemma 5.2. *The* SHORT $\mathfrak{P}$-LCFRS($\mathfrak{P}$) DERIVATION *problem is* in W[1].

Proof. We can restrict ourselves to the case where no nonterminal appears twice in a right-hand side of any rule in G. This is because, e.g., a rule of the form $A \rightarrow f(B, B)$ can be turned into $A \rightarrow f(B, B')$, using a fresh copy B' of B that has the same rules as B (except for having the left-hand side B' rather than B). Note that this modification does not affect the parameter, and increases the size of the grammar only polynomially.

The complete reduction is somewhat lengthy, but the core intuition is very simple. The string is kept the same, and a context-sensitive grammar H is constructed such that $\mathcal{L}(H) = \mathcal{L}(G)$. H simulates G by maintaining a string serialization of the current "configuration" of G, walking through the whole string rewriting the appropriate non-terminal for every rule application in G. A configuration of G can be viewed in this way,

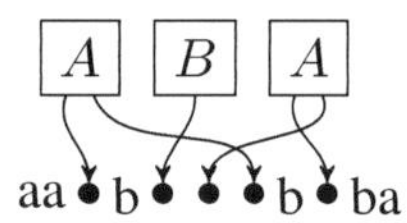

where the derivation has, so far, generated some terminal symbols (the lower-case letters), two instances of the non-terminal A and one instance of B. The configuration keeps track of where the symbols generated by the non-terminals should go in the string, so $\#(A) = 2$, $\#(B) = 1$, and if $(c, d) \in \mathcal{L}_A$ and $e \in \mathcal{L}_B$ this derivation can generate the final string $aacbeddbcba$. These intermediary configurations are in H serialized into strings of nonterminals, with a "nonterminal marker" symbol in each position where a nonterminal is referred to (i.e., H generates a symbol stating "the ith string generated by instance j of the nonterminal A goes here"). H then operates like a Turing machine. A special nonterminal, the rewriting head, picks a rule from G to apply, and walks through the string replacing the nonterminal markers that are affected by that rule. This procedure is then repeated d times.

We start by illustrating the principles of the reduction by an example. Consider the grammar

$$\rhd P_{r_1,1\to 2}X_{S,1,1}\lhd \Longrightarrow \rhd X_{A,1,2}X_{A,2,2}P_{r_1,1\to 2}\lhd \Longrightarrow \rhd X_{A,1,2}X_{A,2,2}R\lhd \Longrightarrow \rhd X_{A,1,2}RX_{A,2,2}\lhd \Longrightarrow \rhd RX_{A,1,2}X_{A,2,2}\lhd \Longrightarrow \rhd P_{r_2,2\to 3}X_{A,1,2}X_{A,2,2}\lhd \stackrel{*}{\Longrightarrow} \rhd X_{A,1,3}X_{B,1,3}X_{A,2,3}B_{2,3}R\lhd \stackrel{*}{\Longrightarrow} \rhd P_{r_2,3\to 4}X_{A,1,3}X_{B,1,3}X_{A,2,3}X_{B,2,3}\lhd \stackrel{*}{\Longrightarrow} \rhd X_{A,1,4}X_{B,1,4}X_{B,1,3}X_{A,2,3}X_{B,2,4}X_{B,2,3}R\lhd \stackrel{*}{\Longrightarrow} \rhd P_{r_6,3\to 1}X_{A,1,4}X_{B,1,4}X_{B,1,3}X_{A,2,3}X_{B,2,4}X_{B,2,3}\lhd \stackrel{*}{\Longrightarrow} \rhd P_{r_3,4\to 1}X_{A,1,4}X_{B,1,4}bX_{A,2,3}X_{B,2,4}b\lhd \stackrel{*}{\Longrightarrow} \rhd P_{r_5,4\to 1}aX_{B,1,4}baX_{B,2,4}b\lhd \stackrel{*}{\Longrightarrow} \rhd aabaabR\lhd \stackrel{*}{\Longrightarrow} aabaab$$

Figure 2: A derivation in the context-sensitive grammar constructed to simulate an LCFRS. All steps in the application of the first rule, $r_1 = S \to f(A)$, are given, the rest is abbreviated.

$G = (\{S, A, B\}, \{a, b\}, F, R, S)$ where F is

$$\{f(x, y) = xy, \quad h_a() = (a, a), \quad h_b() = (b, b), \\ g((x, y), (x', y')) = (xx', yy')\},$$

and R contains the following

$$\begin{array}{ll} r_1 = S \to f(A) & r_2 = A \to g(A, B) \\ r_3 = A \to h_a() & r_4 = A \to h_b() \\ r_5 = B \to h_a() & r_6 = B \to h_b() \end{array}$$

Notice that $\mathcal{L}(G) = \{ww \mid w \in \{a, b\}^+\}$. We now describe how H is constructed by the reduction, after which the more general description follows. A derivation in G starts with the nonterminal S and must then apply r_1. H is constructed to start with the string $\rhd P_{r_1,1\to 2}X_{S,1,1}\lhd$ (all these symbols are nonterminals, H has the same terminal alphabet as G). The symbols $\rhd$ and $\lhd$ mark the beginning and end of the string. The nonterminal $X_{S,1,1}$ is a "nonterminal marker" and denotes the location where the *first* string generated by instance 1 of the nonterminal S is to be placed. Since $\#(S) = 1$ the first string is the only string generated from S. The last subscript, the instance number, is there to differentiate markers belonging to different instances of the same nonterminal. The rewriting head non-deterministically picks an instance number for a round of rewriting (single rule application) from a pool sufficiently large to differentiate between the maximal number of nonterminals (since the rank of G is at most k, no more than k^2 nonterminals can be generated in k rule applications). $P_{r_1,1\to 2}$ is the "rewriting head", the anchor for rule applications. The subscripts on P determines that it will apply the rule r_1, rewriting nonterminal markers corresponding to the left hand side nonterminal of r_1 which have instance number 1. Applying the rule may create new nonterminal markers, all of which get the instance number 2, also determined by the subscript.

That is, the rules for $P_{r_1,i\to j}$ in H will be $P_{r_1,i\to j}X_{S,1,i} \to X_{A,1,j}X_{A,2,j}P_{r_1,i\to j}$, for $i, j \in [2k^2]$, and $P_{r_1,i\to j}x \to xP_{r_1,i\to j}$ for all other $x \neq \lhd$. $P_{r_5,i\to j}X_{B,1,i} \to aP_{r_5,i\to j}$ is another example of a rule corresponding to rule r_5 of G. When a rewriting head hits $\lhd$ it is replaced by a nonterminal R which reverses through the string (with rules of the form $xR \to Rx$ for all $x \neq \rhd$), after which a new rewriting head is non-deterministically picked using one of the rules in $\{\rhd R \to \rhd P_{r,i\to j} \mid r \in R, i, j \in [2k^2]\}$, after which the string is rewritten once more. Finally, there are rules $\rhd \to \varepsilon$, $\lhd \to \varepsilon$ and $R \to \varepsilon$, to remove all nonterminals once rewriting has terminated. A derivation is demonstrated in Figure 2.

By induction on the length of derivations, one can show that $\mathcal{L}(H) = \mathcal{L}(G)$. Now we need to modify the construction slightly to ensure that H can simulate d steps of G in d_H steps.

Limiting steps in G. Construct a SHORT $\mathfrak{P}$-LCFRS($\mathfrak{P}$) DERIVATION instance (G', w, d) from (G, w, d) where G' is such that it *cannot* perform more than d derivation steps. Let

$$N' = \{A_i \mid A \in N, i \in [d]\},$$

and let

$$A_i \to f(B_{j_1}, C_{j_2}, \ldots) \in R'$$

for all $A \to f(B, C, \ldots) \in R$, $i \in [d]$ and $j_1 + j_2 + \cdots = i - 1$. Then $G' = (N', \Sigma, R', S_1)$. This reduction is somewhat heavy-handed, but is in FPT since it leaves k unchanged and each rule is replaced by less than k^k rules (since d and the rank of the grammar are part of the parameter k).

Deferring terminals. A problem in completing the reduction from (G, w, d) to (H, w, d_H) is that the number of terminal symbols G generates is not in its parameter k. For example, G may contain a rule like $A \to a \cdots a$, for an arbitrary number of as. Applying this rule may make the intermediary string H is operating on too long for it to complete rewriting in d_H steps. This can easily be fixed by a polynomial-time rewriting of H. For any rule $w \to w'$ in H such that w' contains at least one terminal, replace every maximal substring $\alpha \in \Sigma^*$ by a new nonterminal T_α, a "terminal place-holder".The rewriting head P and reversal nonterminal R just walk over the place-holders without changing them. Now add the rule $T_\alpha \to \alpha$ for each T_α. For example, where a rewriting head in H might have replaced $X_{A,1,1}$ by $abcX_{B,1,1}baX_{B,2,1}cc$ it will now instead replace it by $T_{abc}X_{B,1,1}T_{ba}X_{B,2,1}T_{cc}$, and can defer replacing the place-holder nonterminals until the end.

Completing the reduction. Now we are ready to put all the pieces together. Given the SHORT $\mathfrak{P}$-LCFRS($\mathfrak{P}$) DERIVATION instance (G, w, d), apply the limiting steps reduction to construct (G', w, d'). Apply the rewriting construction to G to get the context-sensitive grammar H. Now $\mathcal{L}(H)$ equals the language G can generate in d steps. Apply the deferring terminals construction to H to get H'. All that remains is to calculate d_H, the number of steps that H' may take. For an FPT-reduction this number may only depend on the parameter k of (G', w, d'). Picking $d_H = k^5 + 10^3$ is sufficient. Each rule in G' generates less than k nonterminals (since the maximum rank is at most k), each of which will generate at most k markers in the derivation in H' (since the fanout is at most k). The rule may in addition generate $(k+1)k$ terminal place-holders (the k^2 nonterminal markers and string ends separating maximal terminal substrings). After k rule applications, without replacing terminal placeholders, the intermediary string in a derivation in H is less than $k(k^2+(k+1)k)+3$ symbols long. Simulating a rule application in H' entails walking the string twice (forward and then reversing), and k rules are applied, giving $2k(k(k^2+(k+1)k)+3)$ steps. Another $k(k+1)+3$ steps at the end replace the terminal place-holders and remove markers and the rewriting head. Adding things up we arrive at a polynomial of degree 4 that can be rounded up to $k^5 + 10^3$. □

Theorem 5.3. SHORT $\mathfrak{P}$-LCFRS($\mathfrak{P}$) DERIVATION *is* W[1]*-hard.*

Proof. This combines Lemmas 5.1 and 5.2. □

The result of Theorem 5.3 also trivially applies to another natural choice of parameters, the depth of acyclic LCFRS, since they can naturally only take a limited number of derivation steps.

Definition 5.3. A LCFRS is *acyclic* of depth d if d is the smallest integer such that there is a function $\phi : N \to [d]$ such that for all $A \to f(B_1, \ldots, B_n)$ in R and $i \in [n]$ it holds that $\phi(A) < \phi(B_i)$.

Corollary. *The membership problem for acyclic LCFRS where the rank, fan-out, and depth are taken as the parameter is* W[1]*-complete.*

6 Discussion

We have shown that the 1-LCFRS($\mathfrak{P}$)-MEMBERSHIP problem is W[SAT]-hard, but we have no upper bound, except for the trivial XP membership. A conjecture of Pietrzak (2003) may help explain the difficulty of finding such an upper bound. It states that any parameterized problem that has a property that Pietrzak calls *additive* is either in FPT or not in W[P]. Basically, additivity says that any number of instances, sharing a parameter value, can in polynomial time be combined into one big instance, with the same parameter. While 1-LCFRS($\mathfrak{P}$)-MEMBERSHIP is not additive, it has subproblems that are. This means that if Pietrzak's conjecture is true (and FPT $\neq$ W[P]), then 1-LCFRS($\mathfrak{P}$)-MEMBERSHIP cannot belong to W[P].

While our results are mostly intractability results, we see them as a first step towards a more finely grained understanding of the complexity of LCFRS parsing. Ruling out simple parameterization by fan-out or rank as a road towards efficient algorithms lets us focus on other possibilities. Many possible parameterizations remain unexplored. In particular, we conjecture that parameterizing by string length yields FPT membership. In the search for features that can be used in algorithm development, it may also be useful to investigate other formalisms, such as e.g., hypergraph replacement and tree-walking transducers.

Acknowledgments

We acknowledge the support of the Swedish Research Council grant 621-2011-6080.

References

K. A. Abrahamson, R. G. Downey, and M. R. Fellows. 1993. Fixed-parameter intractability II (Extended abstract). In *Proceedings of the 10th Annual Symposium on Theoretical Aspects of Computer Science (STACS'93)*, pages 374–385.

K. A. Abrahamson, R. G. Downey, and M. R. Fellows. 1995. Fixed-parameter tractability and completeness IV: On completeness for W[P] and PSPACE-analogues. *Annals of Pure and Applied Logic*, 73:235–276.

P. Boullier. 2004. Range concatenation grammars. In *New Developments in Parsing Technology*, pages 269–289. Kluwer Academic Publishers.

H. Burden and P. Ljunglöf. 2005. Parsing linear context-free rewriting systems. In *Proceedings of 9th International Workshop on Parsing Technologies*.

R. G. Downey and M. R. Fellows. 1999. *Parameterized Complexity*. Springer-Verlag.

R. G. Downey, M. R. Fellows, B. M. Kapron, M. T. Hallett, and H. T. Wareham. 1994. The parameterized complexity of some problems in logic and linguistics. In *Logical Foundations of Computer Science*, pages 89–100. Springer.

J. Flum and M. Grohe. 2006. *Parameterized Complexity Theory*. Springer-Verlag.

C. Gómez-Rodríguez, M. Kuhlmann, and G. Satta. 2010. Efficient parsing of well-nested linear context-free rewriting systems. In *Human Language Technologies: The 2010 Annual Conference of the North American Chapter of the ACL*, pages 276–284.

A. Joshi. 1985. Tree adjoining grammars: How much context-sensitivity is required to provide reasonable structural descriptions. In *Natural Language Parsing*, pages 206–250. Cambridge University Press.

Y. Kaji, R. Nakanisi, H. Seki, and T. Kasami. 1992. The universal recognition problem for multiple context-free grammars and for linear context-free rewriting systems. *IEICE Transactions on Information and Systems*, E75-D(1):78–88.

K. Pietrzak. 2003. A conjecture on the parameterized hierarchy. Notes on a talk given at Dagstuhl Seminar 03311. Unpublished manuscript.

B. Sagot and G. Satta. 2010. Optimal rank reduction for linear context-free rewriting systems with fan-out two. In *Proceedings of the 48th Annual Meeting of the Association for Computational Linguistics (ACL'10)*, pages 525–533.

G. Satta. 1992. Recognition of linear context-free rewriting systems. In *Proceedings of the 30th Annual Meeting of the Association for Computational Linguistics (ACL'92)*, pages 89–95.

G. Satta. 1998. Trading independent for synchronized parallelism in finite copying parallel rewriting systems. *J. Computer and System Sciences*, 56(1):27–45.

H. Seki, T. Matsumura, M. Fujii, and T. Kasami. 1991. On multiple context-free grammars. *Theoretical Computer Science*, 88(2):191–229.

K. Vijay-Shanker, D. J. Weir, and A. K. Joshi. 1987. Characterizing structural descriptions produced by various grammatical formalisms. In *Proceedings of the 25th Meeting of the Association for Computational Linguists (ACL'87)*, pages 104–111.

Segmenting Temporal Intervals for Tense and Aspect

Tim Fernando
Trinity College Dublin, Ireland
`Tim.Fernando@tcd.ie`

Abstract

Timelines interpreting interval temporal logic formulas are segmented into strings which serve as semantic representations for tense and aspect. The strings have bounded but refinable granularity, suitable for analyzing (im)perfectivity, durativity, telicity, and various relations including branching.

1 Introduction

A sentence in the simple past, such as (1a), uttered at (speech) time S can be pictured as a timeline (1b), describing an event E (Ernest explaining) prior to S.

(1) a. Ernest explained.

b. $\boxed{E}\boxed{S}$ (depicting $E \prec S$)

We can view the event E in (1b) as an unbroken point, wholly to the left of S, $E \prec S$. By contrast, in the timeline (2a) for the progressive (2b), E splits into three boxes, the middle of which contains also a *reference time* R (Reichenbach, 1947).[1]

(2) a. $\boxed{E}\boxed{E,R}\boxed{E}$ (depicting $R \sqsubset E$)

b. Ernest explaining

The relation of R *inside* E, $R \sqsubset E$, breaks E apart, moving us away from conceptualizing E as a point. Indeed, it has become common practice in linguistic semantics since (Bennett and Partee, 1972) to evaluate temporal formulas at intervals, rather than simply points. Interval temporal logics are, however, notoriously more complex than ordinary (pointwise) temporal logics (Halpern and Shoham, 1991; Marcinkowski and Michaliszyn, 2013). That said, for linguistic applications to tense and aspect, the present paper derives strings such as (1b) and (2a) from timelines for interval temporal logic, in effect reducing these timelines to finite models of ordinary temporal logic. This reduction rests on certain assumptions that require explanation and defense.

We begin with temporal formulas, which for the sake of brevity, we hereafter call *fluents*. A fluent such as E, R or S can occur as a whole, as E and S do in (1b), or as segmented, as E does in (2a). We formulate the notions of *whole* and *segmented* model-theoretically in section 2, defining a map $\varphi \mapsto \varphi_\circ$ on fluents φ through which the picture (2a) is sharpened to (3) with $E_\circ$ segmented.

(3) $\boxed{E_\circ}\boxed{E_\circ,R}\boxed{E_\circ}$ (segmented $E_\circ$, whole R)

The map $\varphi \mapsto \varphi_\circ$ is essentially a universal grinder (the right half of an adjoint pair with a universal packager, max)

$$\frac{\text{whole}}{\text{segmented}} \approx \frac{\text{count}}{\text{mass}}$$

pointing to well-known "parallels between the mass-count distinction in nominal systems and the aspectual classification of verbal expressions" (Bach, 1986a). The aspectual classification to which the whole/segmented contrast pertains is that of perfectives and imperfectives

$$\frac{\text{whole}}{\text{segmented}} \approx \frac{\text{perfective}}{\text{imperfective}}$$

as opposed to Aktionsart. A variant of the Aristotle-Ryle-Kenny-Vendler aspectual classes (Dowty, 1979) which can be reduced to durativity and telicity (Comrie, 1976; Moens and Steedman, 1988; Pulman, 1997) is analyzed in section 3 through strings that arise naturally in the investigation of grinding in section 2.

Some restraint on grinding is called for, as the simplest strings are the most coarse-grained. Section 4 enshrines this restraint as a principle, whilst

[1] Boxes are drawn instead of the usual curly braces {, } around the elements of a set to reinforce a reading of (1b) and (2a) as comic strips, with time standing still within a box, but between boxes, progressing from left to right.

accommodating refinements as required. The idea is that strings can be refined by enlarging some contextually supplied set X of (interesting) fluents: the larger X is, the finer the grain becomes. An inverse system of string functions π_X indexed by different finite sets X of fluents is constructed, and applied for an account of relations between strings as well as branching time. The relations here go beyond the familiar order $\prec$ for tense, stretching to the progressive and the perfect, from a variety of perspectives.

2 Segmented versus whole fluents

Fix a set Φ of fluents. Fluents in Φ are interpreted relative to a Φ-*timeline*, a triple $\mathfrak{A} = \langle T, \prec, \models\rangle$ consisting of a linear order $\prec$ on a non-empty set T of (temporal) points, and a binary relation $\models$ between intervals I (over $\prec$) and fluents $\varphi \in \Phi$. An interval is understood here to be a nonempty subset I of T with no holes — i.e. $t \in I$ whenever $t_1 \prec t \prec t_2$ for some pair of points t_1, t_2 in I.[2] $I \models \varphi$ is pronounced "φ holds at I" or "I satisfies φ" (in $\mathfrak{A}$).

A fluent φ is said to be $\mathfrak{A}$-*segmented* if for all intervals I and I' such that $I \cup I'$ is an interval, φ holds at I and at I' precisely if it does at their union

$$I \models \varphi \text{ and } I' \models \varphi \iff I \cup I' \models \varphi.$$

A simple way for a fluent φ to be $\mathfrak{A}$-segmented is by holding at an interval I precisely if it holds at all points of I

$$I \models \varphi \iff (\forall t \in I)\ \{t\} \models \varphi$$

in which case we say φ is $\mathfrak{A}$-*pointed*.[3] A fluent is $\mathfrak{A}$-*singular* if at most one interval satisfies it. Generalizing $\mathfrak{A}$-singular fluents, we call a fluent φ $\mathfrak{A}$-*whole* if for all intervals I and I' such that $I \cup I'$ is an interval,

$$I \models \varphi \text{ and } I' \models \varphi \text{ implies } I = I'.$$

That is, any number of intervals may satisfy a $\mathfrak{A}$-whole fluent so long as no two form an interval. A $\mathfrak{A}$-whole fluent φ defines a *quantized* predicate (Krifka, 1998) insofar as *no* two distinct intervals can satisfy φ if one is a subset of the other. But the ban on pairs of intervals satisfying φ is wider under $\mathfrak{A}$-wholeness. For example, over $T = \{1, 2\}$, a fluent holding at exactly $\{1\}$ and $\{2\}$ is *not* whole, even though $\{\{1\}, \{2\}\}$ is quantized.

$\mathfrak{A}$-wholeness shares half of $\mathfrak{A}$-segmentedness: a fluent φ is $\mathfrak{A}$-*summable* if for all intervals I and I' in $\mathfrak{A}$ such that $I \cup I'$ is an interval,

$$I \models \varphi \text{ and } I' \models \varphi \text{ implies } I \cup I' \models \varphi.$$

Apart from the restriction that $I \cup I'$ is an interval, $\mathfrak{A}$-summability coincides with additivity in (Bach, 1981), illustrated in (4).

(4) Ed slept from 3 to 5pm, Ed slept from 4 to 6pm $\vdash$ Ed slept from 3 to 6pm

The other half of $\mathfrak{A}$-segmentedness (differentiating it from $\mathfrak{A}$-wholeness) is the subinterval property (Bennett and Partee, 1972), enjoyed by states and activities.

(5) Ed slept from 3 to 6 $\vdash$ Ed slept from 3 to 5

A fluent φ is $\mathfrak{A}$-*subinterval-persistent* ($\mathfrak{A}$-sip) if for all intervals I and I' in $\mathfrak{A}$,

$$I \subseteq I' \text{ and } I' \models \varphi \text{ implies } I \models \varphi\ .$$

It is useful to associate with any fluent φ a fluent $\varphi_\circ$ that holds precisely at subintervals of intervals satisfying φ

$$I \models \varphi_\circ \iff (\exists I' \supseteq I)\ I' \models \varphi\ .$$

We say φ is $\mathfrak{A}$-*equivalent to* ψ and write $\varphi \equiv_{\mathfrak{A}} \psi$ if for every interval I,

$$I \models \varphi \iff I \models \psi.$$

Clearly, φ is $\mathfrak{A}$-sip iff $\varphi \equiv_{\mathfrak{A}} \varphi_\circ$. Also, $\varphi_\circ$ is $\mathfrak{A}$-sip and we can say more if φ is $\mathfrak{A}$-whole.

2.1 An adjoint pair

The map $\varphi \mapsto \varphi_\circ$ is one half of a pair for breaking down and building up fluents. To describe the other half, more definitions are helpful. Given a fluent φ and a relation r between intervals, let us form the modal fluent $\langle \mathrm{r}\rangle\varphi$ that holds at an interval r-related to one satisfying φ

$$I \models \langle \mathrm{r}\rangle\varphi \iff (\exists I')\ I \mathrel{\mathrm{r}} I' \text{ and } I' \models \varphi.$$

Note $\varphi_\circ$ is just $\langle\subseteq\rangle\varphi$. Apart from $\subseteq$, other useful examples of relations r between intervals I and I' include *full precedence* $\prec$

$$I \prec I' \iff (\forall t \in I)(\forall t' \in I')\ t \prec t'$$

[2] Not much would be lost were we to take an interval I, as in (Halpern and Shoham, 1991), to be a pair of points t, t' with $t \preceq t'$, or, as in (Allen, 1983), $t \prec t'$.

[3] For finite T, $\mathfrak{A}$-segmented is the same as $\mathfrak{A}$-pointed.

and a relation $\mathtt{m}$ called *meet* in (Allen, 1983) and *abutment* in (Hamblin, 1971).

$$I \,\mathtt{m}\, I' \iff I \prec I' \text{ and } I \cup I' \text{ is an interval.}$$

Now, let $\mathtt{mi}$ be the inverse of $\mathtt{m}$

$$I \,\mathtt{mi}\, I' \iff I' \,\mathtt{m}\, I$$

and max be a function on fluents that maps a fluent ψ to its conjunction with $\neg\langle\mathtt{mi}\rangle\psi$ and $\neg\langle\mathtt{m}\rangle\psi$

$$\max(\psi) \;=\; \psi \wedge \neg\langle\mathtt{mi}\rangle\psi \wedge \neg\langle\mathtt{m}\rangle\psi.$$

Proposition 1.

(a) For all $\mathfrak{A}$-whole φ, $\varphi_\circ$ is $\mathfrak{A}$-segmented and $\varphi \equiv_{\mathfrak{A}} \max(\varphi_\circ)$.

(b) For all $\mathfrak{A}$-segmented ψ, $\max(\psi)$ is $\mathfrak{A}$-whole and $\psi \equiv_{\mathfrak{A}} (\max(\psi))_\circ$.

As to the promised adjunction, let us agree to write $\varphi_{\mathfrak{A}}$ for the set of intervals satisfying φ

$$\varphi_{\mathfrak{A}} \;=\; \{I \mid I \models \varphi\}$$

(so $\varphi \equiv_{\mathfrak{A}} \psi$ iff $\varphi_{\mathfrak{A}} = \psi_{\mathfrak{A}}$) from which we define two pre-orders on fluents

$$\psi \subseteq_{\mathfrak{A}} \psi' \iff \psi_{\mathfrak{A}} \subseteq \psi'_{\mathfrak{A}}$$
$$\varphi \subseteq^{\mathfrak{A}} \varphi' \iff (\forall I \in \varphi_{\mathfrak{A}})(\exists I' \in \varphi'_{\mathfrak{A}})\; I \subseteq I'$$

that apply to $\mathfrak{A}$-segmented fluents ψ and $\mathfrak{A}$-whole fluents φ respectively, for the equivalence

$$\max(\psi) \subseteq^{\mathfrak{A}} \varphi \iff \psi \subseteq_{\mathfrak{A}} \varphi_\circ$$

making max left (lower) adjoint to (of) $\cdot_\circ$.

Next, we turn to linguistic applications and the correspondences

$$\frac{\text{whole}}{\text{segmented}} \approx \frac{\text{count}}{\text{mass}} \approx \frac{\text{perfective}}{\text{imperfective}}.$$

The notion that imperfectives are mass and perfectives count is argued in (Herweg, 1991), building on (Galton, 1984; Galton, 1987) for a concept of *pofective event-type* very close to that of $\mathfrak{A}$-whole fluent above. Perfectives contrast with imperfectives according to (6).

(6) a. viewed from outside, completed, closed

b. viewed from inside, ongoing, open-ended

Towards formalizing (6), let us say an interval I is *inside* an interval I', written $I \sqsubset I'$, if I' extends to the left and also to the right of I

$$I \sqsubset I' \iff (\exists t' \in I')\; \{t'\} \prec I \text{ and } (\exists t'' \in I')\; I \prec \{t''\}$$

(called *during* in (Allen, 1983)). Next, we introduce an $\mathfrak{A}$-whole fluent V for viewpoint to picture a perfective view (6a) of E and an imperfective view (6b) as (7a) and (7b) respectively.

(7) a. $\boxed{\mathrm{V}_\circ}\boxed{\mathrm{E},\mathrm{V}_\circ}\boxed{\mathrm{V}_\circ}$ (depicting $\mathrm{E} \sqsubset \mathrm{V}$)

b. $\boxed{\mathrm{E}_\circ}\boxed{\mathrm{E}_\circ,\mathrm{V}}\boxed{\mathrm{E}_\circ}$ (depicting $\mathrm{V} \sqsubset \mathrm{E}$)

The idea now is to spell out what strings such as (7a) and (7b) mean.

2.2 Segmentations and strings

A *segmentation (seg)* is a sequence $\mathbb{I} = I_1 I_2 \cdots I_n$ of intervals such that

$$I_i \,\mathtt{m}\, I_{i+1} \text{ for } 1 \leq i < n$$

or equivalently,

$$\bigcup_{i=1}^{n} I_i \text{ is an interval, and } I_i \prec I_{i+1} \text{ for } 1 \leq i < n.$$

Given a sequence $\mathbb{I} = I_1 I_2 \cdots I_n$ of intervals and an interval I, we write $\mathbb{I} \nearrow I$ to mean

$$\mathbb{I} \text{ is a seg and } I = \bigcup_{i=1}^{n} I_i,$$

in which case we say $\mathbb{I}$ is a seg(mentation) of I. We extend satisfaction $\models$ to segs $I_1 \cdots I_n$ and strings $\alpha_1 \cdots \alpha_m$ of finite subsets α_i of Φ, requiring that the lengths be the same and that I_i satisfy every fluent in α_i

$$I_1 \cdots I_n \models \alpha_1 \cdots \alpha_m \iff n = m \text{ and } (\forall \varphi \in \alpha_i)\; I_i \models \varphi \text{ for } 1 \leq i \leq n.$$

For example, if E and V are $\mathfrak{A}$-singular (or just $\mathfrak{A}$-whole)

$$(\exists \mathbb{I})\; \mathbb{I} \models \boxed{\mathrm{E}_\circ}\boxed{\mathrm{E}_\circ,\mathrm{V}}\boxed{\mathrm{E}_\circ} \iff (\exists I \models \mathrm{E})\; (\exists J \models \mathrm{V})\; J \sqsubset I.$$

Next, $\mathbb{I} \models s$ extends from a string s to a set L of strings, with *L holding at $\mathbb{I}$* if some string in L does

$$\mathbb{I} \models L \iff (\exists s \in L)\; \mathbb{I} \models s.$$

We then define φ to be $\mathfrak{A}$-*segmentable as* L if an interval I in $\mathfrak{A}$ satisfies φ iff every, or equivalently, some seg of I satisfies L

$$\begin{aligned} I \models \varphi &\iff (\forall \mathbb{I} \nearrow I)\, \mathbb{I} \models L \\ &\iff (\exists \mathbb{I} \nearrow I)\, \mathbb{I} \models L\,. \end{aligned}$$

Proposition 2. *If φ is $\mathfrak{A}$-summable, $\varphi_\circ$ is $\mathfrak{A}$-segmentable as the infinite language*

$$\boxed{\varphi_\circ}^{+} = \boxed{\varphi_\circ} + \boxed{\varphi_\circ}\boxed{\varphi_\circ} + \boxed{\varphi_\circ}\boxed{\varphi_\circ}\boxed{\varphi_\circ} + \cdots$$

of strings $\boxed{\varphi_\circ}^{n}$, $n \geq 1$. Moreover, the following five conditions are pairwise equivalent.

(i) φ *is* $\mathfrak{A}$*-segmented*

(ii) φ *is* $\mathfrak{A}$*-segmentable as* $\boxed{\varphi_\circ}^{+}$

(iii) φ *is* $\mathfrak{A}$*-segmentable as* $\boxed{\varphi}^{+}$

(iv) φ *is* $\mathfrak{A}$*-sip and* $\mathfrak{A}$*-summable*

(v) $\varphi \equiv_{\mathfrak{A}} \max(\varphi)_\circ$.

As for $\mathfrak{A}$-whole fluents, we bound the strings in $\boxed{\varphi_\circ}^{+}$, adding $\neg\langle \mathtt{mi}\rangle\varphi_\circ$ to the initial boxes and $\neg\langle \mathtt{m}\rangle\varphi_\circ$ to the final boxes to form the language

$$\begin{aligned} \mathcal{L}(\varphi) = &\boxed{\varphi_\circ, \neg\langle \mathtt{mi}\rangle\varphi_\circ, \neg\langle \mathtt{m}\rangle\varphi_\circ} + \\ &\boxed{\varphi_\circ, \neg\langle \mathtt{mi}\rangle\varphi_\circ}\boxed{\varphi_\circ}^{*}\boxed{\varphi_\circ, \neg\langle \mathtt{m}\rangle\varphi_\circ}. \end{aligned}$$

Proposition 3. *The following conditions (i)-(iv) are pairwise equivalent.*

(i) φ *is* $\mathfrak{A}$*-whole*

(ii) $\varphi \equiv_{\mathfrak{A}} \max(\varphi_\circ)$

(iii) φ *is* $\mathfrak{A}$*-segmentable as* $\mathcal{L}(\varphi)$

(iv) $\mathbb{I} \models \boxed{\varphi}\boxed{\varphi_\circ} + \boxed{\varphi_\circ}\boxed{\varphi}$ *for no seg* $\mathbb{I}$.

3 Durative and/or telic strings

For any integer $n > 1$, an interval may have a wide variety of segmentations of length n, Propositions 2 and 3 notwithstanding. Even if

$$I \models \mathrm{V} \wedge \langle \sqsupset \rangle \mathrm{E},$$

a seg $I_1 I_2$ of I need not satisfy

$$\boxed{\mathrm{V}_\circ, \langle \sqsupseteq \rangle \mathrm{E}}\boxed{\mathrm{V}_\circ} + \boxed{\mathrm{V}_\circ}\boxed{\mathrm{V}_\circ, \langle \sqsupseteq \rangle \mathrm{E}}$$

(as E may straddle the line between I_1 and I_2), and if E is $\mathfrak{A}$-singular, the string

$$\boxed{\mathrm{V}_\circ}\boxed{\mathrm{E}, \mathrm{V}_\circ}\boxed{\mathrm{V}_\circ}$$

holds in only one out of a possible multitude of segs of I with length 3. The choice of a seg can be a delicate matter. A string of sets of fluents expresses such a choice. The present section links that choice to aspect, stepping from a fluent φ to a set L of strings of finite sets of fluents, with*out* requiring that L hold at every seg of every interval satisfying φ. That is, the account of aspect given below makes essential use of the string representations over and above the fluents from which the strings are formed. Fluents/intervals describe objective matters of fact; strings/segmentations embody, in addition, particular perspectives on these matters.

A concrete linguistic illustration is provided by the notion that some events are *punctual* — i.e., lacking in internal structure. (Comrie, 1976) discusses the example of *cough*, noting that "the inherent punctuality of *cough* would restrict the range of interpretations that can be given to imperfective forms of this verb" to an iterative reading (of a series of coughs), as opposed to a single cough, which he refers to as *semelfactive*. Comrie concedes, however, that, in fact, one can imagine

> a situation where someone is commenting on a slowed down film which incorporates someone's single cough, as for instance in an anatomy lecture: here, it would be quite appropriate for the lecturer to comment on the relevant part of the film *and now the subject is coughing*, even in referring to a single cough, since the single act of coughing has now been extended, and is clearly durative, in that the relevant film sequence lasts for a certain period of time. (page 43)

The earlier contention that *coughing* can only be read iteratively suggests that the intervals spanned by single coughs are too small for our "normal" segmentations. These segmentations consist of intervals too big for "punctual" events, leading to a representation of a φ-semelfactive as $\boxed{\langle \sqsupset \rangle \varphi}$ rather than say, (8), with a middle box $\boxed{\varphi_\circ}$ of internal structure supporting the progressive.

(8) $\boxed{\varphi_\circ, \neg\langle \mathtt{mi}\rangle\varphi_\circ}\boxed{\varphi_\circ}\boxed{\varphi_\circ, \neg\langle \mathtt{m}\rangle\varphi_\circ}$

The special context provided above by an anatomy lecture overturns this restriction, making (8) available after all. The punctual-durative distinction is evidently not cast in stone. But just what is durative? The simple proposal this section explores is that what is durative is a string $\alpha_1\alpha_2\cdots\alpha_n$ of sets α_i of fluents with $n \geq 3$. Between the first box α_1 and the last box α_n is a string $\alpha_2\cdots\alpha_{n-1}$ representing internal structure that, for $n \geq 3$, is non-empty.[4]

Apart from the length n of a string $\alpha_1\cdots\alpha_n$, there is also the matter of what fluents to box in a string, describing the interior as well as the immediate exterior of the situation the string represents. (The string in (8) is just an example to flesh out or otherwise revise.) Of particular relevance to temporal extent are any fluents chosen to mark the boundaries of the situation. An example in (9) is the fluent ψ which makes the string "telic" by appearing in all its boxes negated, except for the rightmost box, which it marks.

(9) $\boxed{\varphi_\circ, \neg\psi}\boxed{\varphi_\circ, \neg\psi}\boxed{\psi}$

Whether or not the intervals described by α_1 and α_n count as part of the situation represented by the string is independent of (10).

(10) a. $\alpha_1\cdots\alpha_n$ is *durative* if it has length $n \geq 3$

b. $\alpha_1\cdots\alpha_n$ is *telic* if the negation of some ψ in α_n appears in α_i for $1 \leq i < n$.

While (10a) says $\alpha_1\cdots\alpha_n$ has internal structure, (10b) says $\alpha_1\cdots\alpha_n$ culminates in some fluent $\psi \in \alpha_n$. (10b) is even more representational than (10a) in that it depends not only on segmenting an interval but on the choice of fluents we put into a string describing that segmentation. As Krifka notes, the telic-atelic distinction lies not "in the nature of the object described, but in the description applied to the object" as

> one and the same event of running can be described by running (i.e. by an atelic predicate) or by running a mile (i.e. a telic, or delimited, predicate)

(Krifka, 1998, page 207).[5] Krifka goes on to locate telicity not in objects but in sets P of objects meeting the condition in (11a), based on a proper part relation $<$ on objects induced by a sum operation $\oplus$ according to (11b).

(11) a. P is *quantized* if there are *no* $x, y \in P$ such that $x < y$

b. $x < y \iff x \neq y$ and $x \oplus y = y$

Under (11), the predicate *run a mile* is quantized, whereas the predicate *run* is not, even though one and the same run may belong to both predicates. But what about *run to the post office*? Surely, the second half of any run to the post office is also a run to the post office. A telic string may fail to be quantized because its left boundary (inception) has not been specified.

3.1 Subsumption and superposition

Some notation from (Fernando, 2004) will come in handy in what follows. Given strings s and s' of sets, we say s *subsumes* s' and write $s \trianglerighteq s'$ if they have the same length and are related componentwise by inclusion

$$\alpha_1\cdots\alpha_n \trianglerighteq \alpha'_1\cdots\alpha'_m \iff n = m \text{ and } \alpha_i \supseteq \alpha'_i \text{ for } 1 \leq i \leq n.$$

For instance,

$$\boxed{\varphi, \neg\psi}\boxed{\varphi, \neg\psi}\boxed{\varphi, \neg\psi}\boxed{\psi} \trianglerighteq \boxed{\varphi}\boxed{\varphi}\boxed{\varphi}\boxed{}.$$

We extend subsumption $\trianglerighteq$ to languages L existentially (just as we did with $\models$)

$$s \trianglerighteq L \iff (\exists s' \in L)\ s \trianglerighteq s'$$

so that a string s is durative iff $s \trianglerighteq \boxed{}\boxed{}\boxed{}^+$ and telic iff $s \trianglerighteq \boxed{\neg\psi}^+\boxed{\psi}$ for some ψ. A binary operation on strings of the same length complementing subsumption $\trianglerighteq$ is *superposition* & obtained by componentwise union

$$\alpha_1\cdots\alpha_n \,\&\, \alpha'_1\cdots\alpha'_n = (\alpha_1 \cup \alpha'_1)\cdots(\alpha_n \cup \alpha'_n).$$

[4] Segmentations of the full linear order T into 2 or 3 intervals are central to the interpretation of *event radicals* in (Galton, 1987). A *formal occurrence* is defined there to be a pair B, A of intervals such that either $AB \nearrow T$ or $AIB \nearrow T$ where I is the complement $T - (A \cup B)$. The intuition is that B is *b*efore, and A *a*fter the situation with temporal extent $T - (A \cup B)$. The first box α_1 and last box α_n of a string $\alpha_1\cdots\alpha_n$ above (with $n \geq 3$) represent final and initial subintervals of B and A, respectively (constituting external structure). The middle bit $\alpha_2\cdots\alpha_{n-1}$ describes a segmentation of $T - (A \cup B)$. Segs generalize formal occurrences, elaborating on internal as well as external structure.

[5] Notice that the condition (10b) for telicity is not met by (8), but by the string

$$\boxed{\varphi_\circ, \neg\langle\mathtt{mi}\rangle\varphi_\circ, \langle\mathtt{m}\rangle\varphi_\circ}\boxed{\varphi_\circ, \langle\mathtt{m}\rangle\varphi_\circ}\boxed{\varphi_\circ, \neg\langle\mathtt{m}\rangle\varphi_\circ}$$

provided $\langle\mathtt{m}\rangle\varphi_\circ$ is understood to be the negation of $\neg\langle\mathtt{m}\rangle\varphi_\circ$. An alternative to leaving ψ existentially quantified in (10b) is to specify the fluent ψ and work with the notion of "culiminating in ψ."

For instance, $\boxed{\varphi}\boxed{\varphi}\boxed{\varphi}$ & $\boxed{\neg\psi}\boxed{\neg\psi}\boxed{\psi}$ $=$ $\boxed{\varphi,\neg\psi}\boxed{\varphi,\neg\psi}\boxed{\varphi,\psi}$ and for strings s and s' of the same length,

$$s \trianglerighteq s' \iff s = s \,\&\, s'$$
$$s \,\&\, s' = \text{least } \trianglerighteq\text{-upper bound of } s \text{ and } s' .$$

We extend & to sets L and L' of strings (of possibly different lengths) by collecting superpositions of strings from L and L' of the same length

$$L \,\&\, L' = \{s \,\&\, s' \mid s \in L,\ s' \in L' \text{ and length}(s)\text{=length}(s')\}$$

(a regular language provided L and L' are (Fernando, 2004)). Notice that

$$\{s\}\&\{s'\} = \{s\&s'\} \quad \text{if length}(s)\text{= length}(s')$$

and the durative strings in L can be obtained by superposing L with $\boxed{}\boxed{}\boxed{}^{+}$

$$L\&\boxed{}\boxed{}\boxed{}^{+} = \{s \in L \mid s \trianglerighteq \boxed{}\boxed{}\boxed{}^{+}\}.$$

3.2 Application to Aktionsart

Semelfactives, activities (= processes), achievements (= culminations) and accomplishments (= culminated processes) are commonly differentiated on the basis of durativity and telicity (Moens and Steedman, 1988; Pulman, 1997).

(12) a. A semelfactive is non-durative and atelic

b. An activity is durative but atelic

c. An achievement is non-durative but telic

d. An accomplishment is telic and durative

Under the present approach based on strings, (12) can be sharpened to (13).

(13) a. A φ-semelfactive $\trianglerighteq \boxed{\langle\supset\rangle\varphi}$

b. A φ-activity $\trianglerighteq \boxed{\varphi}\boxed{\varphi}\boxed{\varphi}^{+}$ (presupposing φ is $\mathfrak{A}$-segmented)

c. A ψ-achievement $\trianglerighteq \boxed{\neg\psi}\boxed{\psi}$

d. An accomplishment built from a φ-activity culminating in ψ

$$\trianglerighteq \boxed{\varphi,\neg\psi}\boxed{\varphi,\neg\psi}\boxed{\varphi,\neg\psi}^{+}\boxed{\psi}$$

(presupposing φ is $\mathfrak{A}$-segmented)

(Bach, 1986a) argues that processes are mass and events are count, raising the question: how does the $\mathfrak{A}$-segmented/whole opposition sit with our account (13) of semelfactives, activities, achievements and accomplishments? Bach's processes are the activities in (13b), represented by the durative strings in the language $\boxed{\varphi}^{+}$ that a $\mathfrak{A}$-segmented fluent φ is $\mathfrak{A}$-segmentable as. Where $\mathfrak{A}$-whole fluents fit in (13) is, however, not immediately obvious. But as pointed out by (Comrie, 1976) for coughs and by (Krifka, 1998) for (mile-long) runs, there is an element of perspective (over and above pure, objective facts) that makes Aktionsart pliable. An achievement may, for instance, be coerced into an accomplishment to interpret the progressive in (14).

(14) The train was arriving when Anna went to order a drink.

A seg II' satisfying an achievement $\boxed{\neg\psi}\boxed{\psi}$ might, for some segmentation $I_1I_2I_3$ of I, be refined to the seg $I_1I_2I_3I'$ satisfying an accomplishment $\boxed{\varphi,\neg\psi}\boxed{\varphi,\neg\psi}\boxed{\varphi,\neg\psi}\boxed{\psi}$ with preparatory process/activity $\boxed{\varphi}\boxed{\varphi}\boxed{\varphi}$, for some $\mathfrak{A}$-segmented φ.

As representations, strings are slippery in a way that fixed pairs $\mathfrak{A}, \mathbb{I}$ are not; a shorter string might describe a larger interval than a longer string does. Strings are not so much finished objects as makeshift representations subject to refinement. So should $\mathfrak{A}$-whole fluents go into these strings? The simplest examples of $\mathfrak{A}$-whole-fluents are $\mathfrak{A}$-singular fluents (harking back to Davidson's events as particulars). Conceptualizing event time at some level of abstraction as an interval is reason enough to form a fluent picking out that interval. And with an $\mathfrak{A}$-singular fluent φ comes the $\mathfrak{A}$-segmented fluent $\varphi_\circ$, and the fluents $\neg\langle\mathtt{mi}\rangle\varphi_\circ$ and $\neg\langle\mathtt{m}\rangle\varphi_\circ$ from which to form the language $\mathcal{L}(\varphi)$ which φ is $\mathfrak{A}$-segmentable as (Proposition 3).

(Dowty, 1979) explores the hypothesis that

> the different aspectual properties of the various kinds of verbs can be explained by postulating a single homogeneous class of predicates – stative predicates – plus three or four sentential operators and connectives

(page 71). A simplified event-based reformulation (15) of the Vendler classes in terms of Dowty's operators DO, BECOME and CAUSE is given in (Rothstein, 2004), page 35.

(15) states $\lambda e.P(e)$

activities $\lambda e.(\text{DO}(P))(e)$

achievements $\lambda e.(\text{BECOME}(P))(e)$

accomplishments $\lambda e.\exists e_1, e_2.[e = e_1 \oplus_S e_2 \wedge (\text{DO}(P))(e_1) \wedge \text{Cul}(e) = e_2]$

Dowty's CAUSE operator is reworked in (15) with a sum operation $\oplus_S$ producing singular entities, and a culmination function Cul. The resulting accomplishment e is the sum $e_1 \oplus_S e_2$ of its preparatory process (activity) e_1 and culmination e_2. To bring (13) in line with (15), we put

$$\begin{aligned} \text{DO}(P) &\approx \boxed{P}\boxed{P}\boxed{P}^+ \\ \text{BECOME}(P) &\approx \boxed{\neg P}\boxed{P} \end{aligned}$$

and require that P be $\mathfrak{A}$-segmented. Defining

$$\begin{aligned} \text{du}(L) &= L \& \boxed{}\boxed{}^+ \\ \text{cu}(L, \psi) &= (L \& \boxed{\neg\psi}^+)\boxed{\psi} \end{aligned}$$

yields

$$\begin{aligned} \boxed{P}\boxed{P}\boxed{P}^+ &= \text{du}(\boxed{P}^+) \\ \boxed{\neg P}\boxed{P} &= \text{cu}(\boxed{}, P) \end{aligned}$$

and for accomplishments as culiminated activities,

$$\begin{aligned} \text{cu}(\text{du}(\boxed{\varphi}^+), \psi) &= \boxed{\varphi, \neg\psi}\boxed{\varphi, \neg\psi}\boxed{\varphi, \neg\psi}^+\boxed{\psi} \\ &= \text{du}(\boxed{\varphi, \neg\psi})\boxed{\psi}. \end{aligned}$$

Left out of (13) are the states in (15), which can be compared to $\mathfrak{A}$-segmented fluents in the present framework. As noted in (Dowty, 1986), one might also require that stative fluents be inertial, for which see (Fernando, 2008).

4 Desegmenting and branching time

Why segment an interval? The two reasons given above are (1) to get a handle on durativity and telicity, and (2) to unwind an interval fluent such as $\text{E} \wedge \langle \sqsupset \rangle \text{R}$ to a string $\boxed{\text{E}_\circ}\boxed{\text{E}_\circ,\text{R}}\boxed{\text{E}_\circ}$ interpreted against segmentations (i.e. finite timelines). Neither reason justifies grinding indefinitely. The thrust of the present section is to leave segs as coarse as possible, segmenting only if necessary, leading to a notion of time that may branch.

4.1 Desegmenting via π

Quantifying the model $\mathfrak{A}$ out of the notion of $\mathfrak{A}$-segmentability and weakening the connection between intervals and segs, let us agree that a language L *depicts* φ if for all models $\mathfrak{A}$, L is $\mathfrak{A}$-satisfiable precisely if φ is

$$(\exists \text{ seg } \mathbb{I})\ \mathbb{I} \models L \iff (\exists \text{ interval } I)\ I \models \varphi.$$

Trivially, $\boxed{\varphi}$ depicts φ, but there are more interesting examples. Unwinding the modal operator $\langle \succ \rangle$ and conjunction $\wedge$ in the fluent $\text{S} \wedge \langle \succ \rangle \varphi$,

$$\boxed{\varphi}\boxed{\text{S}} + \boxed{\varphi}\boxed{}\boxed{\text{S}} \text{ depicts } \text{S} \wedge \langle \succ \rangle \varphi.$$

The language $\boxed{\varphi}\boxed{\text{S}} + \boxed{\varphi}\boxed{}\boxed{\text{S}}$ reduces the infinite language $\boxed{}^*\boxed{\varphi}\boxed{}^*\boxed{\text{S}}\boxed{}^*$ depicting $\text{S} \wedge \langle \succ \rangle \varphi$ to two strings. This reduction illustrates the possibility that under suitable assumptions on a language L depicting φ, the strings in L can be simplified in two ways:

(w1) any initial or final empty boxes can be stripped off, and

(w2) all repeating blocks α^n (for $n \geq 1$) of a box α can be compressed to α.

More precisely, we implement (w1) by a function *unpad* defined on strings s by

$$\mathit{unpad}(s) = \begin{cases} \mathit{unpad}(s') & \text{if } s = \boxed{}s' \text{ or} \\ & \quad \text{else } s = s'\boxed{} \\ s & \text{otherwise} \end{cases}$$

so that $\mathit{unpad}(s)$ neither begins nor ends with $\boxed{}$. For (w2), all blocks α^{n+1} in s are compressed in $bc(s)$ to α

$$bc(s) = \begin{cases} bc(\alpha s') & \text{if } s = \alpha\alpha s' \\ \alpha\ bc(\beta s') & \text{if } s = \alpha\beta s' \text{ with} \\ & \quad \alpha \neq \beta \\ s & \text{otherwise} \end{cases}$$

so that if $bc(s) = \alpha_1 \cdots \alpha_n$ then $\alpha_i \neq \alpha_{i+1}$ for i from 1 to $n-1$. We then compose bc with *unpad* for π

$$\pi(s) = \mathit{unpad}(bc(s)).$$

One can check that

$$\{\pi(s) \mid s \in \boxed{}^*\boxed{\varphi}\boxed{}^*\boxed{\text{S}}\boxed{}^*\} = \boxed{\varphi}\boxed{\text{S}} + \boxed{\varphi}\boxed{}\boxed{\text{S}}.$$

Clearly, $\pi(s)$ is never longer than s, and $\pi(s) = \pi(\pi(s))$ for all strings s.

As for the "suitable assumptions" on L under which L can be reduced to $\{\pi(s) | s \in L\}$, it is helpful to consider the fluent $\mathrm{R} \wedge \langle\sqsubset\rangle\varphi$. Can we unwind $\langle\sqsubset\rangle$ in $\boxed{\mathrm{R}, \langle\sqsubset\rangle\varphi}$? Assuming $\varphi_\circ$ is $\mathfrak{A}$-summable for all models $\mathfrak{A}$,

$$\boxed{\varphi_\circ}\boxed{\mathrm{R},\varphi_\circ}\boxed{\varphi_\circ} \text{ depicts } \mathrm{R} \wedge \langle\sqsubset\rangle\varphi.$$

Now, let us call a string $s = \alpha_1 \cdots \alpha_n$ of sets α_i of fluents *$\mathfrak{A}$-reducible* if every fluent appearing in two consecutive string positions $\alpha_i\alpha_{i+1}$ in s (for some $1 \leq i < n$) is $\mathfrak{A}$-summable. (For example, $\boxed{\varphi_\circ}\boxed{\mathrm{R},\varphi_\circ}\boxed{\varphi_\circ}$ is $\mathfrak{A}$-reducible, provided $\varphi_\circ$ is $\mathfrak{A}$-summable.) Let us say a seg $\mathbb{I}$ *refines* a seg $I_1 \cdots I_n$ if for all i from 1 to n, I_i is the union of some subsequence of $\mathbb{I}$.

Proposition 4. *For any $\mathfrak{A}$-reducible string s, every seg $\mathbb{I}$ that satisfies s refines some seg $\mathbb{I}'$ that satisfies $\pi(s)$. Consequently, if for all $s \in L$, s is $\mathfrak{A}$-reducible and $\pi(s) \in L$, then L is $\mathfrak{A}$-satisfiable iff $\{\pi(s) | s \in L\}$ is*

$$(\exists \text{ seg } \mathbb{I})\ \mathbb{I} \models L \iff (\exists \text{ seg } \mathbb{I})\ \mathbb{I} \models \{\pi(s) \mid s \in L\}.$$

4.2 Relativizing π to a finite set X of fluents

Next, we fix a notion of bounded granularity through a finite set X of fluents of interest, which we can expand to refine granularity or contract to coarsen granularity. An instructive example for orientation is the representation of a calendar year of twelve months as the string

$$s_{mo} = \boxed{\mathsf{Jan}}\boxed{\mathsf{Feb}}\boxed{\mathsf{Mar}} \cdots \boxed{\mathsf{Dec}}$$

of length 12, or, were we also interested in days d1,d2...,d31, the string

$$s_{mo,dy} = \boxed{\mathsf{Jan,d1}}\boxed{\mathsf{Jan,d2}} \cdots \boxed{\mathsf{Jan,d31}} \boxed{\mathsf{Feb,d1}} \cdots \boxed{\mathsf{Dec,d31}}$$

of length 365 (for a non-leap year). Unlike the points in the real line $\mathbb{R}$, a box can split, as $\boxed{\mathsf{Jan}}$ in s_{mo} does (30 times) to $\boxed{\mathsf{Jan,d1}}\boxed{\mathsf{Jan,d2}} \cdots \boxed{\mathsf{Jan,d31}}$ in $s_{mo,dy}$, on introducing days d1, d2,..., d31 into the picture. Reversing direction and generalizing from mo = {Jan,Feb,...Dec} to any set X of fluents, we define the function ρ_X on strings (of sets) to componentwise intersect with X

$$\rho_X(\alpha_1 \cdots \alpha_n) = (\alpha_1 \cap X) \cdots (\alpha_n \cap X)$$

throwing out non-X's from each box (keeping only the elements of X) so that

$$\rho_{mo}(s_{mo,dy}) = \boxed{\mathsf{Jan}}^{31}\boxed{\mathsf{Feb}}^{28} \cdots \boxed{\mathsf{Dec}}^{31}.$$

Next, we compose ρ_X and π for the function π_X mapping a string s of sets to

$$\pi_X(s) = \pi(\rho_X(s)) = \mathit{unpad}(\mathit{bc}(\rho_X(s)))$$

so that for example,

$$\pi_{mo}(s_{mo,dy}) = \pi(\boxed{\mathsf{Jan}}^{31}\boxed{\mathsf{Feb}}^{28} \cdots \boxed{\mathsf{Dec}}^{31}) = s_{mo}$$

and

$$\pi_{\{\mathrm{E}_\circ\}}(\boxed{\mathrm{E}_\circ}\boxed{\mathrm{R},\mathrm{E}_\circ}\boxed{\mathrm{E}_\circ}) = \pi(\boxed{\mathrm{E}_\circ}\boxed{\mathrm{E}_\circ}\boxed{\mathrm{E}_\circ}) = \boxed{\mathrm{E}_\circ}.$$

In general, a description s_X of granularity X can be refined to one $s_{X'}$ of granularity $X' \supseteq X$ provided π_X maps $s_{X'}$ to s_X. More precisely, given some large set Φ of fluents, let $\mathit{Fin}(\Phi)$ be the set of finite subsets of Φ. A function f with domain $\mathit{Fin}(\Phi)$ mapping $X \in \mathit{Fin}(\Phi)$ to a string $f(X)$ over the alphabet 2^X is said to be *π-consistent* if whenever $X \subseteq X' \in \mathit{Fin}(\Phi)$,

$$f(X) = \pi_X(f(X')).$$

Let us write $\mathfrak{IL}_\pi(\Phi)$ for the set of all π-consistent functions. "IL" here stands not for intensional logic but for inverse limit — to be precise, the *inverse limit of* the restrictions of π_X to $(2^{X'})^*$ for $X \subseteq X' \in \mathit{Fin}(\Phi)$ (all computable by finite-state transducers). That said, $\mathfrak{IL}_\pi(\Phi)$ is intensional: time branches under the relation $\prec_\Phi$ between $f, f' \in \mathfrak{IL}_\pi(\Phi)$ given by

$$f \prec_\Phi f' \iff f \neq f' \text{ and } (\forall X \in \mathit{Fin}(\Phi))\ f(X) \text{ is a prefix of } f'(X)$$

(where s is a *prefix of* s' if $s' = s\hat{s}$ for some possibly empty string $\hat{s}$). The intuition is that a temporal moment comes with its past, and that an $f \in \mathfrak{IL}_\pi(\Phi)$ encodes the moment that is X-approximated, for each $X \in \mathit{Fin}(\Phi)$, by the last box in $f(X)$, with past given by the remainder of $f(X)$ (leading to that box). More precisely, $\prec_\Phi$ is tree-like in the sense of (Dowty, 1979).

Proposition 5. *$\prec_\Phi$ is transitive and left linear: for every $f \in \mathfrak{IL}(\Phi)$,*

$$(\forall f_1 \prec_\Phi f)(\forall f_2 \prec_\Phi f)\ f_1 \prec_\Phi f_2 \text{ or } f_2 \prec_\Phi f_1 \text{ or } f_1 = f_2.$$

Moreover, no element of $\mathfrak{IL}_\pi(\Phi)$ is $\prec_\Phi$-maximal.

Maximal chains, called *histories* in (Dowty, 1979), figure prominently in possible worlds semantics. While we can pick one out in $\mathfrak{IL}_\pi(\Phi)$ to represent an actual history, it is far from obvious what significance maximal $\prec_\Phi$-chains have in the present framework, which is closer in spirit to *situation semantics* (Bawise and Perry, 1983), updated in (Cooper, 2005; Ginzburg, 2005).

Tha handbook chapter (Thomason, 1984) opens with the declaration

> Physics should have helped us to realise that a temporal theory of a phenomenon X is, in general, more than a simple combination of two components: the statics of X and the ordered set of temporal instants. The case in which all functions from times to world-states are allowed is uninteresting; there are too many such functions, and the theory has not begun until we have begun to restrict them. And often the principles that emerge from the interaction of time with the phenomena seem new and surprising.

For a non-empty set W of worlds, and a linearly ordered set T of time instants, Thomason compares $T \times W$-frames, not unlike that in (Montague, 1973), unfavorably to tree-like frames, of which $\prec_\Phi$ above is an example, when paired with a $\subseteq$-maximal $\prec_\Phi$-chain. The crudeness of the cartesian product $\times$ aside, one may ask where T comes from, as Bach pointedly does in page 69 of (Bach, 1981), to say nothing of W. The answer from $\mathfrak{IL}_\pi(\Phi)$ involves strings formed from fluents. The projective system $(\pi_X)_{X \in Fin(\Phi)}$ gives for every finite subset X of Φ, a choice of X-approximations in $(2^X)^*$, including for $X = \{e, e'\}$ with $e \neq e'$, 13 strings s_r corresponding to the Allen interval relations r between intervals e and e' (Allen, 1983); see Table 1 (Fernando, 2012). Under the projections π_X, these strings are most naturally viewed as *indices* for evaluating an expression φ as an *extension* or *denotation*, as prescribed by Carnap-Montague *intensions* (Fernando, 2011). In (Bach, 1986b), an event type such as KISSING induces a function EXT(KISSING) that maps histories to subparts that are temporal manifestations of KISSING, treating input histories as indices and output manifestations as extensions. Under the current framework, EXT(KISSING) can for any $X \in Fin(\Phi)$, be given as a binary relation between strings in $(2^X)^*$ that X-approximate indices and extensions.

$r \in$ **Allen**	$s_r \in (2^{\{e,e'\}})^+$
$e = e'$	$\boxed{e,e'}$
e s e'	$\boxed{e,e'}\boxed{e'}$
e si e'	$\boxed{e,e'}\boxed{e}$
e f e'	$\boxed{e'}\boxed{e,e'}$
e fi e'	$\boxed{e}\boxed{e,e'}$
e d e'	$\boxed{e'}\boxed{e,e'}\boxed{e'}$
e di e'	$\boxed{e}\boxed{e,e'}\boxed{e}$
e o e'	$\boxed{e}\boxed{e,e'}\boxed{e'}$
e oi e'	$\boxed{e'}\boxed{e,e'}\boxed{e}$
e m e'	$\boxed{e}\boxed{e'}$
$e < e'$	$\boxed{e}\boxed{}\boxed{e'}$
e mi e'	$\boxed{e'}\boxed{e}$
$e > e'$	$\boxed{e'}\boxed{}\boxed{e}$

Table 1: The Allen relations in $(2^{\{e,e'\}})^+$

5 Conclusion

Segmentations arise naturally in the view from (Klein, 2009) that

> The expression of time in natural languages relates a *clause-internal temporal structure* to a *clause-external temporal structure*. The latter may shrink to a single interval, for example, the time at which the sentence is uttered; but this is just a special case. The clause-internal temporal structure may also be very simple – it may be reduced to a single interval without any further differentiation, the 'time of the situation'; but if this ever happens, it is only a borderline case. As a rule, the clause-internal struc-

ture is much more complex. (page 75)

The simplest case described by the passage is illustrated by the picture (16) of the clause-internal event (or situation) time E preceding the clause-external speech (utterance) time S.

(16) $\boxed{E|S} + \boxed{E| |S}$ depicting $E \wedge \langle \prec \rangle S$

Slightly more complicated is the picture (3) of event time E with R inside it.

(3) $\boxed{E_\circ | E_\circ,R | E_\circ}$ (segmented $E_\circ$, whole R)

Whereas E in (16) is unbroken and whole, the "differentiation" in (3) puts E through a universal grinder $\cdot_\circ$ described in section 2, alongside notions of $\mathfrak{A}$-whole and $\mathfrak{A}$-segmented fluents, the contrast between which surfaces in pairs such as (17) and (18).

(17) Ernest was explaining $\not\vdash$ Ernest explained

(18) Ernest was laughing $\vdash$ Ernest laughed

The non-entailment (17) is clear from (19).

(19) Ernest was explaining when he was made to stop.

To extract a rigorous account of (17) versus (18) from the assumption that explaining is whole and laughing is segmented (as fluents) would require stepping beyond lexical/internal aspect (considered in sections 2 and 3 above) to grammatical/external aspect, hinted at in (3), as well as tense. Some details compatible with the present approach can be found in (Fernando, 2008).[6] Suffice it to say that additional temporal parameters from tense and aspect enlarge the set X of fluents that, under the inverse limit $\mathfrak{IL}_\pi(\Phi)$ in section 4, refines granularity. While we have taken pains to show how to interpret a string of subsets of Φ in a segmentation (essentially, a finite, ordered partition of an interval from a Φ-timeline), no Φ-timeline is used to define $\mathfrak{IL}_\pi(\Phi)$, resulting in a notion of time that branches (away from any single segmentation or timeline). There is sure to be junk in $\mathfrak{IL}_\pi(\Phi)$ to throw out; but what use tense and aspect might have for timelines not represented in $\mathfrak{IL}_\pi(\Phi)$, I fail to see (apart from linking temporality up with other linguistic mechanisms such as quantification). Work on tense and aspect has led to extensions of ordinary temporal logic in three directions.

(20) a. addition of temporal parameters (e.g. R)

b. expansion of points to intervals

c. recognition of events and states

Stringing together finite sets of fluents, we attend to (20c) in sections 2 and 3 above, and to (20a) in section 4, putting the distinction (20b) between points and intervals down to the set X of fluents under consideration.[7]

[6]An alternative would be to follow along (Galton, 1984; Galton, 1987). There are likely to be many ways to fill in the details. In the case of the perfect, for instance, the basic approach outlined here is, as far as I can tell, neutral between extended now accounts (Pancheva, 2003) augmented with (7)

(7) a. $\boxed{V_\circ | E,V_\circ | V_\circ}$ (depicting $E \sqsubset V$)

b. $\boxed{E_\circ | E_\circ,V | E_\circ}$ (depicting $V \sqsubset E$)

and consequent-state approaches (Moens and Steedman, 1988; Kamp and Reyle, 1993; Pulman, 1997) that might be augmented with inertia (Dowty, 1986) and forces (Fernando, 2008).

[7]*Added in haste*. The literature on interval temporal logic is vast, and the present paper has doubtless failed to do it justice. In particular, (Nishimura, 1980) and (Moszkowski, 1986) deserve to be mentioned properly in this paper, which I hope to do in a revision.

References

James F. Allen. 1983 Maintaining knowledge about temporal intervals. *Communications of the ACM*, 26:832–843.

Emmon Bach. 1981. On time, tense and aspect: an essay in English metaphysics. In Peter Cole, editor, *Radical Pragmatics*, pages 63 – 81. Academic Press.

Emmon Bach. 1986a. The algebra of events. *Linguistics and Philosophy*, 9:5–16.

Emmon Bach. 1986b. Natural language metaphysics. In R. Barcan Marcus, G.J.W. Dorn, and P. Weingartner, editors, *Logic, Methodology and Philosophy of Science VII*, pages 573 – 595. Elsevier.

Jon Bawise and John Perry. 1983. *Situations and Attitudes*. Bradford, MIT Press.

Michael Bennett and Barbara Partee. 1972. Toward the logic of tense and aspect in English. Indiana University Linguistics Club, Bloomington, IN.

Bernard Comrie. 1976. *Aspect*. Cambridge University Press.

Robin Cooper. 2005. Austinian truth, attitudes and type theory. *Research on Language and Computation*, 3(4):333–362.

David R. Dowty. 1979. *Word Meaning and Montague Grammar*. Reidel, Dordrecht.

David R. Dowty. 1986. The effects of aspectual class on the temporal structure of discourse: semantics or pragmatics? *Linguistics and Philosophy*, 9(1):37–61.

Tim Fernando. 2004. A finite-state approach to events in natural language semantics. *Journal of Logic and Computation*, 14(1):79–92.

Tim Fernando. 2008. Branching from inertia worlds. *J. Semantics*, 25(3):321–344.

Tim Fernando. 2011. Regular relations for temporal propositions. *Natural Language Engineering*, 17(2):163–184.

Tim Fernando. 2012. A finite-state temporal ontology and event-intervals. Proceedings of the 10th International Workshop on Finite State Methods and Natural Language Processing, pages 80–89, Donostia/San Sebastian (ACL archive).

Antony Galton. 1984. *The logic of aspect: an axiomatic approach*. Clarendon Press.

Antony Galton. 1987. The logic of occurrence. In *Temporal Logics and Their Applications*, pages 169–196. Academic Press.

Jonathan Ginzburg. 2005. Situation semantics: the ontological balance sheet. *Research on Language and Computation*, 3(4):363–389.

Joseph Halpern and Yoav Shoham. 1991. A propositional modal logic of time intervals. *Journal of the Association for Computing Machinery*, 38(4):935–962.

Charles L. Hamblin. 1971. Instants and intervals, *Studium Generale*, 24:127–134.

Michael Herweg. 1991. A critical examination of two classical approaches to aspect. *J. Semantics*, 8:363–402.

Hans Kamp and Uwe Reyle. 1993. *From Discourse to Logic*. Kluwer.

Wolfgang Klein. 2009. How time is encoded. In W. Klein and P. Li, editors, *The Expression of Time*, pages 39–81. Mouton De Gruyter.

Manfred Krifka. 1998. The origins of telicity. In S. Rothstein, editor, *Events and Grammar*, pages 197–235. Kluwer.

Jerzy Marcinkowski and Jakub Michaliszyn. 2013. The undecidability of the logic of subintervals. *Fundamenta Informaticae* 20:124.

Marc Moens and Mark Steedman. 1988. Temporal ontology and temporal reference. *Computational Linguistics*, 14(2):15–28.

Richard Montague. 1973. The proper treatment of quantification in ordinary English. In K.J.J. Hintikka, J.M.E. Moravcsik, and P. Suppes, editors, *Approaches to Natural Language*, pages 221–42. D. Reidel, Dordrecht.

Ben Moszkowski. 1986. *Executing Temporal Logic Programs*. Cambridge University Press.

Hirokazu Nishimura. 1980. Interval Logics with Applications to Study of Tense and Aspect in English. Publ. Research Institute for Mathematical Sciences, Kyoto University 16:417-459.

Roumyana Pancheva. 2003. The aspectual makeup of Perfect participles and the interpretations of the Perfect. In A. Alexiadou and M. Rathert and A. von Stechow, editors, *Perfect Explorations*, pages 277–306. Mouton de Gruyter.

Stephen G. Pulman. 1997. Aspectual shift as type coercion. *Transactions of the Philological Society*, 95(2):279–317.

Hans Reichenbach. 1947. *Elements of Symbolic Logic*. MacMillan Company, NY.

Susan Rothstein. 2004. *Structuring Events: A Study in the Semantics of Lexical Aspect*. Blackwell.

Richmond Thomason. 1984. Combinations of tense and modality. In D. Gabbay and F. Guenthner, editors, *Handbook of Philosophical Logic*, volume II, pages 135–165. Reidel.

The Frobenius Anatomy of Relative Pronouns

Stephen Clark
University of Cambridge
Computer Laboratory
sc609@cam.ac.uk

Bob Coecke
University of Oxford
Dept. of Computer Science
coecke@cs.ox.ac.uk

Mehrnoosh Sadrzadeh
Queen Mary, University of London
School of Electronic
Engineering and Computer Science
mehrnoosh.sadrzadeh@eecs.qmul.ac.uk

Abstract

This paper develops a compositional vector-based semantics of relative pronouns within a categorical framework. Frobenius algebras are used to formalise the operations required to model the semantics of relative pronouns, including passing information between the relative clause and the modified noun phrase, as well as copying, combining, and discarding parts of the relative clause. We develop two instantiations of the abstract semantics, one based on a truth-theoretic approach and one based on corpus statistics.

1 Introduction

Ordered algebraic structures and sequent calculi have been used extensively in Computer Science and Mathematical Logic. They have also been used to formalise and reason about natural language. Lambek (1958) used the ordered algebra of residuated monoids to model grammatical types, their juxtapositions and reductions. Relational words such as verbs have implicative types and are modelled using the residuals to the monoid multiplication. Later, Lambek (1999) simplified these algebras in favour of *pregroups*. Here, there are no binary residual operations, but each element of the algebra has a left and a right residual.

In terms of semantics, pregroups do not naturally lend themselves to a model-theoretic treatment (Montague, 1974). However, pregroups are suited to a radically different treatment of semantics, namely distributional semantics (Schütze, 1998). Distributional semantics uses vector spaces based on contextual co-occurrences to model the meanings of words. Coecke et al. (2010) show how a *compositional* semantics can be developed within a vector-based framework, by exploiting the fact that vector spaces with linear maps and pregroups both have a compact closed categorical structure (Kelly and Laplaza, 1980; Preller and Lambek, 2007). Some initial attempts at implementation include Grefenstette and Sadrzadeh (2011a) and Grefenstette and Sadrzadeh (2011b).

One problem with the distributional approach is that it is difficult to see how the meanings of some words — e.g. logical words such as *and, or,* and relative pronouns such as *who, which, that, whose* — can be modelled contextually. Our focus in this paper is on relative pronouns in the distributional compositional setting.

The difficulty with pronouns is that the contexts in which they occur do not seem to provide a suitable representation of their meanings: pronouns tend to occur with a great many nouns and verbs. Hence, if one applies the contextual co-occurrence methods of distributional semantics to them, the result will be a set of dense vectors which do not discriminate between different meanings. The current state-of-the-art in compositional distributional semantics either adopts a simple method to obtain a vector for a sequence of words, such as adding or mutliplying the contextual vectors of the words (Mitchell and Lapata, 2008), or, based on the grammatical structure, builds linear maps for some words and applies these to the vector representations of the other words in the string (Baroni and Zamparelli, 2010; Grefenstette and Sadrzadeh, 2011a). Neither of these approaches produce vectors which provide a good representation for the meanings of relative clauses.

In the grammar-based approach, one has to assign a linear map to the relative pronoun, for instance a map f as follows:

$$\overrightarrow{\text{men who like Mary}} = f(\overrightarrow{\text{men}}, \overrightarrow{\text{like Mary}})$$

However, it is not clear what this map should be. Ideally, we do not want it to depend on the frequency of the co-occurrence of the relative pronoun with the relevant basis vectors. But both of the above mentioned approaches rely heavily on the information provided by a corpus to build their linear maps. The work of Baroni and Zamparelli (2010) uses linear regression and approximates the context vectors of phrases in which the target word has occurred, and the work of Grefenstette and Sadrzadeh (2011a) uses the sum of Kronecker products of the arguments of the target word across the corpus.

The semantics we develop for relative pronouns and clauses uses the general operations of a Frobenius algebra over vector spaces (Coecke et al., 2008) and the structural categorical morphisms of vector spaces. We do not rely on the co-occurrence frequencies of the pronouns in a corpus and only take into account the structural roles of the pronouns in the meaning of the clauses. The computations of the algebra and vector spaces are depicted using string diagrams (Joyal and Street, 1991), which depict the interactions that occur among the words of a sentence. In the particular case of relative clauses, they visualise the role of the relative pronoun in passing information between the clause and the modified noun phrase, as well as copying, combining, and even discarding parts of the relative clause.

We develop two instantiations of the abstract semantics, one based on a truth-theoretic approach, and one based on corpus statistics, where for the latter the categorical operations are instantiated as matrix multiplication and vector component-wise multiplication. As a result, we will obtain the following for the meaning of a subject relative clause:

$$\overrightarrow{\text{men who like Mary}} = \overrightarrow{\text{men}} \odot (\overline{\text{love}} \times \overrightarrow{\text{Mary}})$$

The rest of the paper introduces the categorical framework, including the formal definitions relevant to the use of Frobenius algebras, and then shows how these structures can be used to model relative pronouns within the compositional vector-based setting.

2 Compact Closed Categories and Frobenius Algebras

This section briefly reviews compact closed categories and Frobenius algebras. For a formal presentation, see (Kelly and Laplaza, 1980; Kock, 2003; Baez and Dolan, 1995), and for an informal introduction see Coecke and Paquette (2008).

A compact closed category has objects A, B; morphisms $f: A \to B$; a monoidal tensor $A \otimes B$ that has a unit I; and for each object A two objects A^r and A^l together with the following morphisms:

$$A \otimes A^r \xrightarrow{\epsilon^r_A} I \xrightarrow{\eta^r_A} A^r \otimes A$$
$$A^l \otimes A \xrightarrow{\epsilon^l_A} I \xrightarrow{\eta^l_A} A \otimes A^l$$

These morphisms satisfy the following equalities, sometimes referred to as the *yanking* equalities, where 1_A is the identity morphism on object A:

$$(1_A \otimes \epsilon^l_A) \circ (\eta^l_A \otimes 1_A) = 1_A$$
$$(\epsilon^r_A \otimes 1_A) \circ (1_A \otimes \eta^r_A) = 1_A$$
$$(\epsilon^l_A \otimes 1_A) \circ (1_{A^l} \otimes \eta^l_A) = 1_{A^l}$$
$$(1_{A^r} \otimes \epsilon^r_A) \circ (\eta^r_A \otimes 1_{A^r}) = 1_{A^r}$$

A pregroup is a partial order compact closed category, which we refer to as *Preg*. This means that the objects of *Preg* are elements of a partially ordered monoid, and between any two objects $p, q \in$ *Preg* there exists a morphism of type $p \to q$ iff $p \leq q$. Compositions of morphisms are obtained by transitivity and the identities by reflexivity of the partial order. The tensor of the category is the monoid multiplication, and the epsilon and eta maps are as follows:

$$\epsilon^r_p = p \cdot p^r \leq 1 \qquad \eta^r_p = 1 \leq p^r \cdot p$$
$$\epsilon^l_p = p^l \cdot p \leq 1 \qquad \eta^l_p = 1 \leq p \cdot p^l$$

Finite dimensional vector spaces and linear maps also form a compact closed category, which we refer to as *FVect*. Finite dimensional vector spaces V, W are objects of this category; linear maps $f: V \to W$ are its morphisms with composition being the composition of linear maps. The tensor product $V \otimes W$ is the linear algebraic tensor product, whose unit is the scalar field of vector spaces; in our case this is the field of reals $\mathbb{R}$. As opposed to the tensor product in *Preg*, the tensor between vector spaces is symmetric; hence we have a natural isomorphism $V \otimes W \cong W \otimes V$. As a result of the symmetry of the tensor, the two adjoints reduce to one and we obtain the following isomorphism:

$$V^l \cong V^r \cong V^*$$

where V^* is the dual of V. When the basis vectors of the vector spaces are fixed, it is further the case that the following isomorphism holds as well:

$$V^* \cong V$$

Elements of vector spaces, i.e. vectors, are represented by morphisms from the unit of tensor to their corresponding vector space; that is $\overrightarrow{v} \in V$ is represented by the morphism $\mathbb{R} \xrightarrow{\overrightarrow{v}} V$; by linearity this morphism is uniquely defined when setting $1 \mapsto \overrightarrow{v}$.

Given a basis $\{r_i\}_i$ for a vector space V, the epsilon maps are given by the inner product extended by linearity; i.e. we have:

$$\epsilon^l = \epsilon^r : V^* \otimes V \to \mathbb{R}$$

given by:

$$\sum_{ij} c_{ij}\ \psi_i \otimes \phi_j \quad \mapsto \quad \sum_{ij} c_{ij} \langle \psi_i \mid \phi_j \rangle$$

Similarly, eta maps are defined as follows:

$$\eta^l = \eta^r : \mathbb{R} \to V \otimes V^*$$

and are given by:

$$1 \mapsto \sum_i r_i \otimes r_i$$

A Frobenius algebra in a monoidal category $(\mathcal{C}, \otimes, I)$ is a tuple $(X, \Delta, \iota, \mu, \zeta)$ where, for X an object of $\mathcal{C}$, the triple (X, Δ, ι) is an internal comonoid; i.e. the following are coassociative and counital morphisms of $\mathcal{C}$:

$$\Delta \colon X \to X \otimes X \qquad\qquad \iota \colon X \to I$$

Moreover (X, μ, ζ) is an internal monoid; i.e. the following are associative and unital morphisms:

$$\mu \colon X \otimes X \to X \qquad\qquad \zeta \colon I \to X$$

And finally the Δ and μ morphisms satisfy the following *Frobenius condition*:

$$(\mu \otimes 1_X) \circ (1_X \otimes \Delta) \;=\; \Delta \circ \mu \;=\; (1_X \otimes \mu) \circ (\Delta \otimes 1_X)$$

Informally, the comultiplication Δ decomposes the information contained in one object into two objects, and the multiplication μ combines the information of two objects into one.

Frobenius algebras were originally introduced in the context of representation theorems for group theory (Frobenius, 1903). Since then, they have found applications in other fields of mathematics and physics, e.g. in topological quantum field theory (Kock, 2003). The above general categorical definition is due to Carboni and Walters (1987). In what follows, we use Frobenius algebras that characterise vector space bases (Coecke et al., 2008).

In the category of finite dimensional vector spaces and linear maps *FVect*, any vector space V with a fixed basis $\{\overrightarrow{v_i}\}_i$ has a Frobenius algebra over it, explicitly given by:

$$\Delta :: \overrightarrow{v_i} \mapsto \overrightarrow{v_i} \otimes \overrightarrow{v_i} \qquad\qquad \iota :: \overrightarrow{v_i} \mapsto 1$$
$$\mu :: \overrightarrow{v_i} \otimes \overrightarrow{v_j} \mapsto \delta_{ij} \overrightarrow{v_i} \qquad\qquad \zeta :: 1 \mapsto \textstyle\sum_i \overrightarrow{v_i}$$

where δ_{ij} is the Kronecker delta.

Frobenius algebras over vector spaces with orthonormal bases are moreover *isometric* and *commutative*. A commutative Frobenius Algebra satisfies the following two conditions for $\sigma : X \otimes Y \to Y \otimes X$, the symmetry morphism of $(\mathcal{C}, \otimes, I)$:

$$\sigma \circ \Delta = \Delta \qquad\qquad \mu \circ \sigma = \mu$$

An isometric Frobenius Algebra is one that satisfies the following axiom:

$$\mu \circ \Delta = 1$$

The vector spaces of distributional models have fixed orthonormal bases; hence they have isometric commutative Frobenius algebras over them.

The comultiplication Δ of an isometric commutative Frobenius Algebra over a vector space encodes vectors of lower dimensions into vectors of higher dimensional tensor spaces; this operation is referred to as *copying*. In linear algebraic terms, $\Delta(\overrightarrow{v}) \in V \otimes V$ is a diagonal matrix whose diagonal elements are weights of $\overrightarrow{v} \in V$. The corresponding multiplication μ encodes vectors of higher dimensional tensor spaces into lower dimensional spaces; this operation is referred to as *combining*. For $\overrightarrow{w} \in V \otimes V$, we have that $\mu(\overrightarrow{w}) \in V$ is a vector consisting only of the diagonal elements of $\overrightarrow{w}$.

As a concrete example, take V to be a two dimensional space with basis $\{\overrightarrow{v_1}, \overrightarrow{v_2}\}$; then the basis of $V \otimes V$ is $\{\overrightarrow{v_1} \otimes \overrightarrow{v_1}, \overrightarrow{v_1} \otimes \overrightarrow{v_2}, \overrightarrow{v_2} \otimes \overrightarrow{v_1}, \overrightarrow{v_2} \otimes \overrightarrow{v_2}\}$. For a vector $v = a\overrightarrow{v_1} + b\overrightarrow{n_2}$ in V we have:

$$\Delta(v) = \Delta \begin{pmatrix} a \\ b \end{pmatrix} = \begin{pmatrix} a & 0 \\ 0 & b \end{pmatrix} = a\, \overrightarrow{v_1} \otimes \overrightarrow{v_1} + b\, \overrightarrow{v_2} \otimes \overrightarrow{v_2}$$

And for a matrix $w = a\,\overrightarrow{v_1} \otimes \overrightarrow{v_1} + b\,\overrightarrow{v_1} \otimes \overrightarrow{v_2} + c\,\overrightarrow{v_2} \otimes \overrightarrow{v_1} + d\,\overrightarrow{v_2} \otimes \overrightarrow{v_2}$ in $V \otimes V$, we have:

$$\mu(w) = \mu \begin{pmatrix} a & b \\ c & d \end{pmatrix} = \begin{pmatrix} a \\ d \end{pmatrix} = a\,\overrightarrow{v_1} + d\,\overrightarrow{v_2}$$

3 String Diagrams

The framework of compact closed categories and Frobenius algebras comes with a complete diagrammatic calculus that visualises derivations, and which also simplifies the categorical and vector space computations. Morphisms are depicted by boxes and objects by lines, representing their identity morphisms. For instance a morphism $f\colon A \to B$, and an object A with the identity arrow $1_A\colon A \to A$, are depicted as follows:

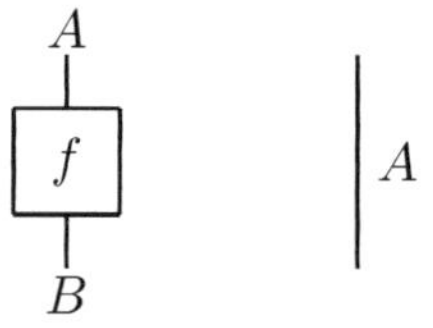

The tensor products of the objects and morphisms are depicted by juxtaposing their diagrams side by side, whereas compositions of morphisms are depicted by putting one on top of the other; for instance the object $A \otimes B$, and the morphisms $f \otimes g$ and $f \circ h$, for $f\colon A \to B, g\colon C \to D$, and $h\colon B \to C$, are depicted as follows:

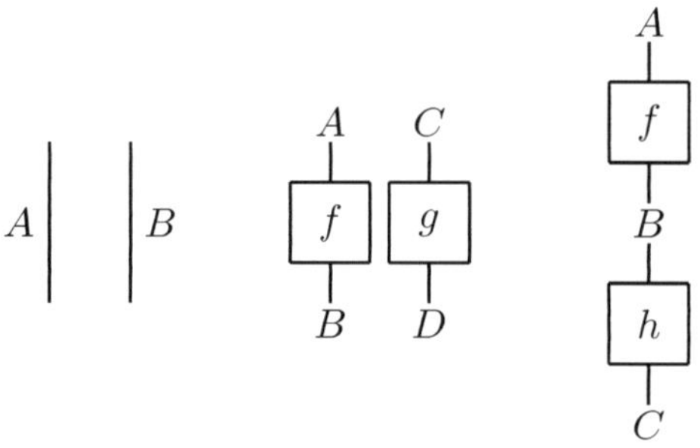

The ϵ maps are depicted by cups, η maps by caps, and yanking by their composition and straightening of the strings. For instance, the diagrams for $\epsilon^l\colon A^l \otimes A \to I$, $\eta\colon I \to A \otimes A^l$ and $(\epsilon^l \otimes 1_A) \circ (1_A \otimes \eta^l) = 1_A$ are as follows:

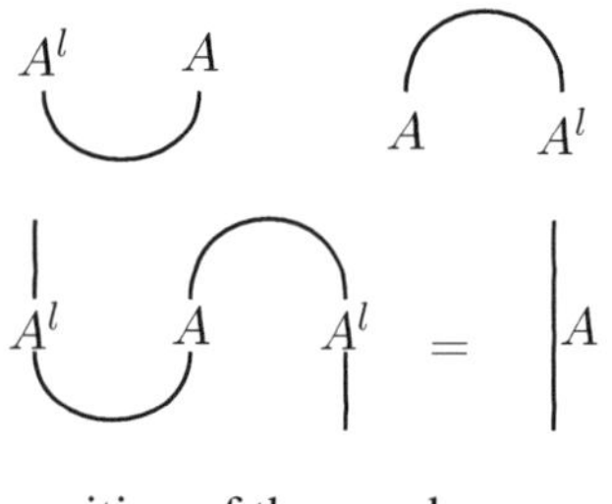

The composition of the ϵ and η maps with other morphisms is depicted as before, that is by juxtaposing them one above the other. For instance the diagrams for the compositions $(1_{B^l} \otimes f) \circ \epsilon^l$ and $\eta^l \circ (1_{A^l} \otimes f)$ are as follows:

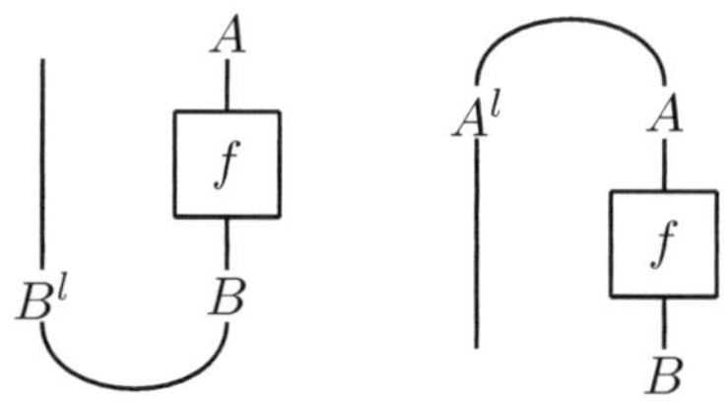

As for Frobenius algebras, the diagrams for the monoid and comonoid morphisms are as follows:

with the Frobenius condition being depicted as:

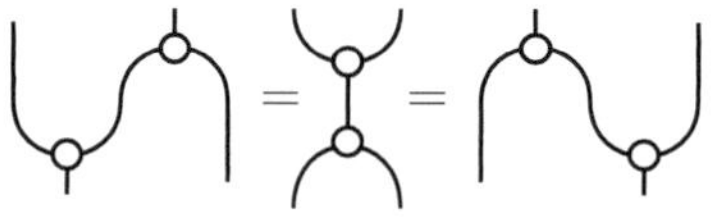

The defining axioms guarantee that any picture depicting a Frobenius computation can be reduced to a normal form that only depends on the number of input and output strings of the nodes, independent of the topology. These normal forms can be simplified to so-called 'spiders':

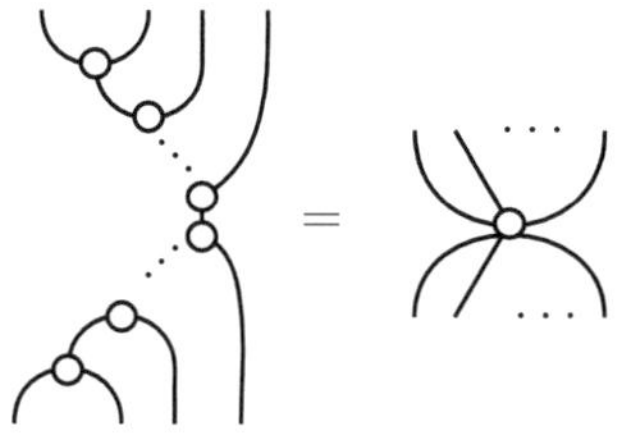

In the category *FVect*, apart from spaces V, W, which are objects of the category, we also have vectors $\overrightarrow{v}, \overrightarrow{w}$. These are depicted by their representing morphisms and as triangles with a number of strings emanating from them. The number of strings of a triangle denote the tensor rank of the vector; for instance, the diagrams for $\overrightarrow{v} \in V$, $\overrightarrow{v'} \in V \otimes W$, and $\overrightarrow{v''} \in V \otimes W \otimes Z$ are as follows:

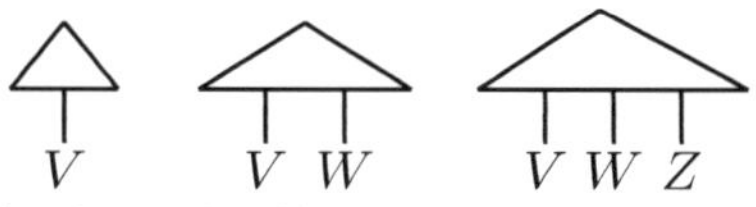

Application of a linear map to a vector is depicted using composition of their corresponding morphisms. For instance, for $f\colon V \to W$ and $\overrightarrow{v} \in V$, the application $f(\overrightarrow{v})$ is depicted by the composition $I \xrightarrow{\overrightarrow{v}} V \xrightarrow{f} W$.

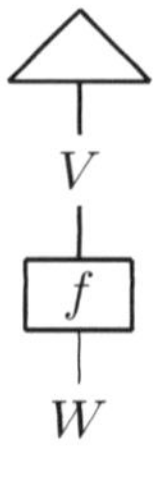

Applications of the Frobenius maps to vectors are depicted in a similar fashion; for instance $\mu(\overrightarrow{v} \otimes \overrightarrow{v})$ is the composition $I \otimes I \xrightarrow{\overrightarrow{v} \otimes \overrightarrow{v}} V \otimes V \xrightarrow{\mu} V$ and $\iota(\overrightarrow{v})$ is the composition $I \xrightarrow{\overrightarrow{v}} V \xrightarrow{\iota} I$, depicted as follows:

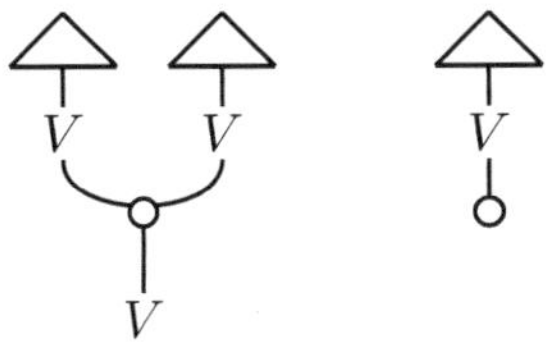

4 Vector Space Interpretations

The grammatical structure of a language is encoded in the category *Preg*: objects are grammatical types (assigned to words of the language) and morphisms are grammatical reductions (encoding the grammatical formation rules of the language). For instance, the grammatical structure of the sentence "Men love Mary" is encoded in the assignment of types n to the noun phrases "men" and "Mary" and $n^r \otimes s \otimes n^l$ to the verb "love", and in the reduction map $\epsilon_n^l \otimes 1_s \otimes \epsilon_n^r$. The application of this reduction map to the tensor product of the word types in the sentence results in the type s:

$$(\epsilon_n^l \otimes 1_s \otimes \epsilon_n^r)(n \otimes (n^r \otimes s \otimes n^l) \otimes n) \to s$$

To each reduction map corresponds a string diagram that depicts the structure of reduction:

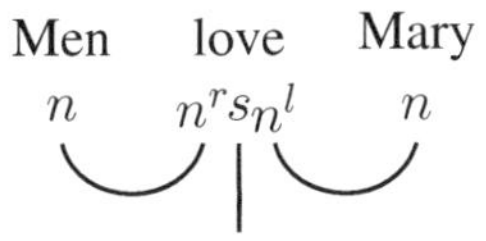

In Coecke et al. (2010) the pregroup types and reductions are interpreted as vector spaces and linear maps, achieved via a homomorphic mapping from *Preg* to *FVect*. Categorically speaking, this map is a strongly monoidal functor:

$$F\colon \textit{Preg} \to \textit{FVect}$$

It assigns vector spaces to the basic types as follows:

$$F(1) = I \qquad F(n) = N \qquad F(s) = S$$

and to the compound types by monoidality as follows; for x, y objects of *Preg*:

$$F(x \otimes y) = F(x) \otimes F(y)$$

Monoidal functors preserve the compact structure; that is the following holds:

$$F(x^l) = F(x^r) = F(x)^*$$

For instance, the interpretation of a transitive verb is computed as follows:

$$F(n^r \otimes s \otimes n^l) = F(n^r) \otimes F(s) \otimes F(n^l) = F(n)^* \otimes F(s) \otimes F(n)^* = N \otimes S \otimes N$$

This interpretation means that the meaning vector of a transitive verb is a vector in $N \otimes S \otimes N$.

The pregroup reductions, i.e. the partial order morphisms of *Preg*, are interpreted as linear maps: whenever $p \leq q$ in *Preg*, we have a linear map $f_{\leq}\colon F(p) \to F(q)$. The ϵ and η maps of *Preg* are interpreted as the ϵ and η maps of *FVect*. For instance, the pregroup reduction of a transitive verb sentence is computed as follows:

$$F(\epsilon_n^r \otimes 1_s \otimes \epsilon_n^r) = F(\epsilon_n^r) \otimes F(1_s) \otimes F(\epsilon_n^l) = F(\epsilon_n)^* \otimes F(1_s) \otimes F(\epsilon_n)^* = \epsilon_N \otimes 1_S \otimes \epsilon_N$$

The distributional meaning of a sentence is obtained by applying the interpretation of the pregroup reduction of the sentence to the tensor product of the distributional meanings of the words in the sentence. For instance, the distributional meaning of "Men love Mary" is as follows:

$$F(\epsilon_n^r \otimes 1_s \otimes \epsilon_n^l)(\overrightarrow{\textit{Men}} \otimes \overrightarrow{\textit{love}} \otimes \overrightarrow{\textit{Mary}})$$

This meaning is depictable via the following string diagram:

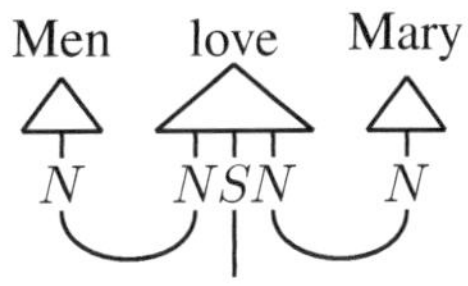

The next section applies these techniques to the distributional interpretation of pronouns. The interpretations are defined using: ϵ maps, for application of the semantics of one word to another; η maps, to pass information around by bridging intermediate words; and Frobenius operations, for copying and combining the noun vectors and discarding the sentence vectors.

5 Modelling Relative Pronouns

In this paper we focus on the subject and object relative pronouns, *who(m)*, *which* and *that*. Examples of noun phrases with subject relative pronouns are "men who love Mary", "dog which ate cats". Examples of noun phrases with object relative pronouns are "men whom Mary loves", "book that John read". In the final example, "book" is the head noun, modified by the relative clause "that John read". The intuition behind the use of Frobenius algebras to model such cases is the following. In "book that John read", the relative clause acts on the noun (modifies it) via the relative pronoun, which passes information from the clause to the noun. The relative clause is then discarded, and the modified noun is returned. Frobenius algebras provide the machinery for all of these operations.

The pregroup types of the relative pronouns are as follows:

$$n^r n s^l n \quad \text{(subject)}$$
$$n^r n n^{ll} s^l \quad \text{(object)}$$

These types result in the following reductions:

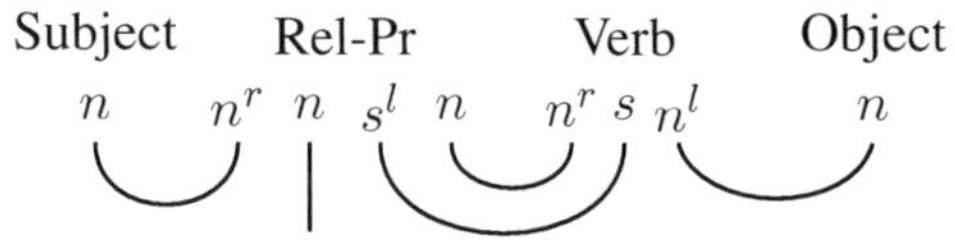

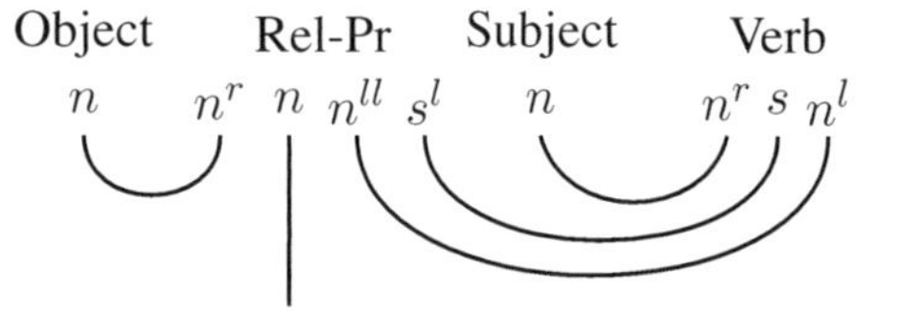

The meaning spaces of these pronouns are computed using the mechanism described above:

$$\begin{aligned} F(n^r n s^l n) &= F(n^r) \otimes F(n) \otimes F(s^l) \otimes F(n) \\ &= N \otimes N \otimes S \otimes N \\ F(n^r n n^{ll} s^l) &= F(n^r) \otimes F(n) \otimes F(n^{ll}) \otimes F(s^l) \\ &= N \otimes N \otimes N \otimes S \end{aligned}$$

The semantic roles that these pronouns play are reflected in their categorical vector space meanings, depicted as follows:

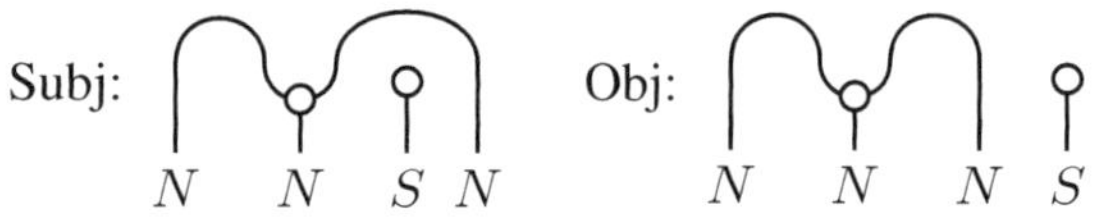

with the following corresponding morphisms:

Subj: $(1_N \otimes \mu_N \otimes \zeta_S \otimes 1_N) \circ (\eta_N \otimes \eta_N)$
Obj: $(1_N \otimes \mu_N \otimes 1_N \otimes \zeta_S) \circ (\eta_N \otimes \eta_N)$

The diagram of the meaning vector of the subject relative clause interacting with the head noun is as follows:

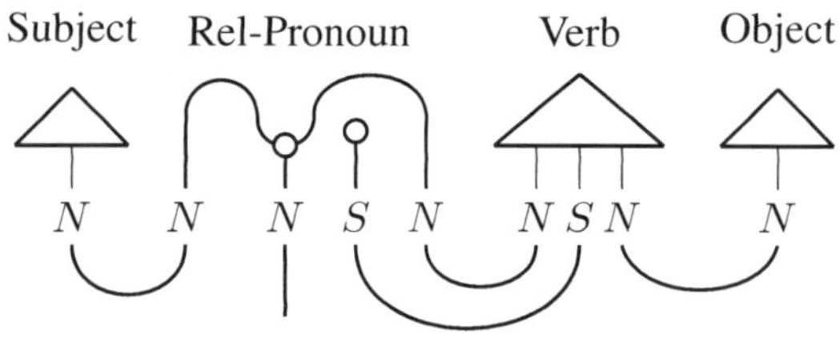

The diagram for the object relative clause is:

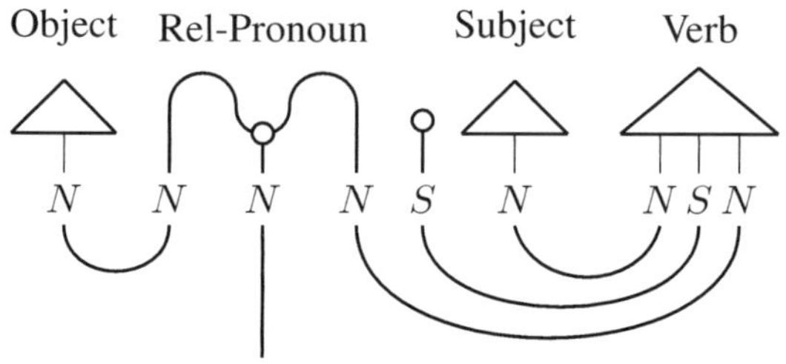

These diagrams depict the flow of information in a relative clause and the semantic role of its relative pronoun, which 1) passes information from the clause to the head noun via the η maps; 2) acts on the noun via the μ map; 3) discards the clause via the ζ map; and 4) returns the modified noun via 1_N. The ϵ maps pass the information of the subject and object nouns to the verb and to the relative pronoun to be acted on. Note that there are two different flows of information in these clauses: the ones that come from the grammatical structure and are depicted by ϵ maps (at the bottom of the diagrams), and the ones that come from the semantic role of the pronoun and are depicted by η maps (at the top of the diagrams).

The normal forms of these diagrams are:

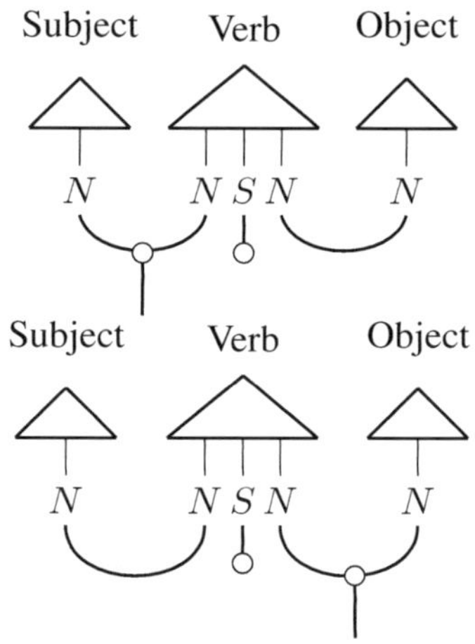

Symbolically, they correspond to the following morphisms:

$$(\mu_N \otimes \iota_S \otimes \epsilon_N)\left(\overrightarrow{\text{Subject}} \otimes \overrightarrow{\text{Verb}} \otimes \overrightarrow{\text{Object}}\right)$$
$$(\epsilon_N \otimes \iota_S \otimes \mu_N)\left(\overrightarrow{\text{Subject}} \otimes \overrightarrow{\text{Verb}} \otimes \overrightarrow{\text{Object}}\right)$$

The simplified normal forms will become useful in practice when calculating vectors for such cases.

6 Vector Space Instantiations

In this section we demonstrate the effect of the Frobenius operations using two example instantiations. The first — which is designed perhaps as a theoretical example rather than a suggestion for implementation — is a truth-theoretic account, similar to Coecke et al. (2010) but also allowing for degrees of truth. The second is based on the concrete implementation of Grefenstette and Sadrzadeh (2011a).

6.1 Degrees of Truth

Take N to be the vector space spanned by a set of individuals $\{\overrightarrow{n}_i\}_i$ that are mutually orthogonal. For example, $\overrightarrow{n}_1$ represents the individual Mary, $\overrightarrow{n}_{25}$ represents Roger the dog, $\overrightarrow{n}_{10}$ represents John, and so on. A sum of basis vectors in this space represents a common noun; e.g. $\overrightarrow{man} = \sum_i \overrightarrow{n}_i$, where i ranges over the basis vectors denoting men. We take S to be the one dimensional space spanned by the single vector $\overrightarrow{1}$. The unit vector spanning S represents truth value 1, the zero vector represents truth value 0, and the intermediate vectors represent degrees of truth.

A transitive verb w, which is a vector in the space $N \otimes S \otimes N$, is represented as follows:

$$\overline{w} := \sum_{ij} \overrightarrow{n}_i \otimes (\alpha_{ij} \overrightarrow{1}) \otimes \overrightarrow{n}_j$$

if $\overrightarrow{n}_i$ w's $\overrightarrow{n}_j$ with degree α_{ij}, for all i, j.

Further, since S is one-dimensional with its only basis vector being $\overrightarrow{1}$, the transitive verb can be represented by the following element of $N \otimes N$:

$$\sum_{ij} \alpha_{ij} \overrightarrow{n}_i \otimes \overrightarrow{n}_j \quad \text{if} \quad \overrightarrow{n}_i \ w\text{'s}\ \overrightarrow{n}_j \text{ with degree } \alpha_{ij}$$

Restricting to either $\alpha_{ij} = 1$ or $\alpha_{ij} = 0$ provides a 0/1 meaning, i.e. either $\overrightarrow{n}_i$ w's $\overrightarrow{n}_j$ or not. Letting α_{ij} range over the interval $[0, 1]$ enables us to represent degrees as well as limiting cases of truth and falsity. For example, the verb "love", denoted by $\overline{love}$, is represented by:

$$\sum_{ij} \alpha_{ij} \overrightarrow{n}_i \otimes \overrightarrow{n}_j \quad \text{if} \quad \overrightarrow{n}_i \text{ loves } \overrightarrow{n}_j \text{with degree } \alpha_{ij}$$

If we take α_{ij} to be 1 or 0, from the above we obtain the following:

$$\sum_{(i,j) \in R_{love}} \overrightarrow{n}_i \otimes \overrightarrow{n}_j$$

where R_{love} is the set of all pairs (i, j) such that $\overrightarrow{n}_i$ loves $\overrightarrow{n}_j$.

Note that, with this definition, the sentence space has already been discarded, and so for this

$$\overrightarrow{\text{Subject who Verb Object}} :=$$
$$(\mu_N \otimes \epsilon_N)\left(\overrightarrow{\text{Subject}} \otimes \overrightarrow{\text{Verb}} \otimes \overrightarrow{\text{Object}}\right) =$$
$$(\mu_N \otimes \epsilon_N)\left(\sum_{k \in K} \overrightarrow{n}_k \otimes (\sum_{ij} \alpha_{ij} \overrightarrow{n}_i \otimes \overrightarrow{n}_j) \otimes \sum_{l \in L} \overrightarrow{n}_l\right)$$
$$= \sum_{ij, k \in K, l \in L} \alpha_{ij} \mu_N(\overrightarrow{n}_k \otimes \overrightarrow{n}_i) \otimes \epsilon_N(\overrightarrow{n}_j \otimes \overrightarrow{n}_l)$$
$$= \sum_{ij, k \in K, l \in L} \alpha_{ij} \delta_{ki} \overrightarrow{n}_i \delta_{jl}$$
$$= \sum_{k \in K, l \in L} \alpha_{kl} \overrightarrow{n}_k$$

Figure 1: Meaning computation with a subject relative pronoun

instantiation the ι map, which is the part of the relative pronoun interpretation designed to discard the relative clause after it has acted on the head noun, is not required.

For common nouns $\overrightarrow{\text{Subject}} = \sum_{k \in K} \overrightarrow{n}_k$ and $\overrightarrow{\text{Object}} = \sum_{l \in L} \overrightarrow{n}_l$, where k and l range over the sets of basis vectors representing the respective common nouns, the truth-theoretic meaning of a noun phrase modified by a subject relative clause is computed as in Figure 1. The result is highly intuitive, namely the sum of the subject individuals weighted by the degree with which they have acted on the object individuals via the verb. A similar computation, with the difference that the μ and ϵ maps are swapped, provides the truth-theoretic semantics of the object relative clause:

$$\sum_{k \in K, l \in L} \alpha_{kl} \overrightarrow{n}_l$$

The calculation and final outcome is best understood with an example.

Now only consider truth values 0 and 1. Consider the noun phrase with object relative clause "men whom Mary loves" and take N to be the vector space spanned by the set of all people; then the males form a subspace of this space, where the basis vectors of this subspace, i.e. men, are denoted by $\overrightarrow{m}_l$, where l ranges over the set of men which we denote by M. We set "Mary" to be the individual $\overrightarrow{f}_1$, "men" to be the common noun $\sum_{l \in M} \overrightarrow{m}_l$,

$$
\begin{aligned}
&\overrightarrow{\text{men whom Mary loves}} := \\
&(\epsilon_N \otimes \mu_N)\left(\overrightarrow{f}_1 \otimes \Big(\sum_{(i,j)\in R_{love}} \overrightarrow{f}_i \otimes \overrightarrow{m}_j\Big) \otimes \sum_{l\in M} \overrightarrow{m}_l\right) \\
&= \sum_{l\in M,(i,j)\in R_{love}} \epsilon_N(\overrightarrow{f}_1 \otimes \overrightarrow{f}_i) \otimes \mu(\overrightarrow{m}_j \otimes \overrightarrow{m}_l) \\
&= \sum_{l\in M,(i,j)\in R_{love}} \delta_{1i}\delta_{jl}\overrightarrow{m}_j \\
&= \sum_{(1,j)\in R_{love}|j\in M} \overrightarrow{m}_j
\end{aligned}
$$

Figure 2: Meaning computation for example object relative clause

and "love" to be as follows:

$$\sum_{(i,j)\in R_{love}} \overrightarrow{f}_i \otimes \overrightarrow{m}_j$$

The vector corresponding to the meaning of "men whom Mary loves" is computed as in Figure 2. The result is the sum of the men basis vectors which are also loved by Mary.

The second example involves degrees of truth. Suppose we have two females Mary $\overrightarrow{f}_1$ and Jane $\overrightarrow{f}_2$ and four men $\overrightarrow{m}_1, \overrightarrow{m}_2, \overrightarrow{m}_3, \overrightarrow{m}_4$. Mary loves $\overrightarrow{m}_1$ with degree 1/4 and $\overrightarrow{m}_2$ with degree 1/2; Jane loves $\overrightarrow{m}_3$ with degree 1/5; and $\overrightarrow{m}_4$ is not loved. In this situation, we have:

$$R_{love} = \{(1,1),(1,2),(2,3)\}$$

and the verb love is represented by:

$$1/4(\overrightarrow{f}_1 \otimes \overrightarrow{m}_1) + 1/2(\overrightarrow{f}_1 \otimes \overrightarrow{m}_2) + 1/5(\overrightarrow{f}_2 \otimes \overrightarrow{m}_3)$$

The meaning of "men whom Mary loves" is computed by substituting an $\alpha_{1,j}$ in the last line of Figure 2, resulting in the men whom Mary loves together with the degrees that she loves them:

$$\sum_{(1,j)\in R_{love}|j\in M} \alpha_{1j}\overrightarrow{m}_j = 1/4\overrightarrow{m}_1 + 1/2\overrightarrow{m}_2$$

"men whom women love" is computed as follows, where W is the set of women:

$$
\begin{aligned}
&\sum_{k\in W,l\in M,(i,j)\in R_{love}} \alpha_{ij}\epsilon_N(\overrightarrow{f}_k \otimes \overrightarrow{f}_i) \otimes \mu(\overrightarrow{m}_j \otimes \overrightarrow{m}_l) \\
&= \sum_{k\in W,l\in M,(i,j)\in R_{love}} \alpha_{ij}\delta_{ki}\delta_{jl}\overrightarrow{m}_j \\
&= \sum_{(i,j)\in R_{love}|i\in W,j\in M} \alpha_{ij}\overrightarrow{m}_j \\
&= 1/4\overrightarrow{m}_1 + 1/2\overrightarrow{m}_2 + 1/5\overrightarrow{m}_3
\end{aligned}
$$

The result is the men loved by Mary or Jane together with the degrees to which they are loved.

6.2 A Concrete Instantiation

In the model of Grefenstette and Sadrzadeh (2011a), the meaning of a verb is taken to be "the degree to which the verb relates properties of its subjects to properties of its object". Clark (2013) provides some examples showing how this is an intuitive defintion for a transitive verb in the categorical framework. This degree is computed by forming the sum of the tensor products of the subjects and objects of the verb across a corpus, where w ranges over instances of the verb:

$$\overline{\text{verb}} = \sum_w (\overrightarrow{\text{sbj}} \otimes \overrightarrow{\text{obj}})_w$$

Denote the vector space of nouns by N; the above is a matrix in $N \otimes N$, depicted by a two-legged triangle as follows:

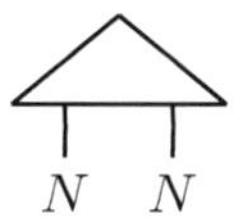

The verbs of this model do not have a sentence dimension; hence no information needs to be discarded when they are used in our setting, and so no ι map appears in the diagram of the relative clause. Inserting the above diagram in the diagrams of the normal forms results in the following for the subject relative clause (the object case is similar):

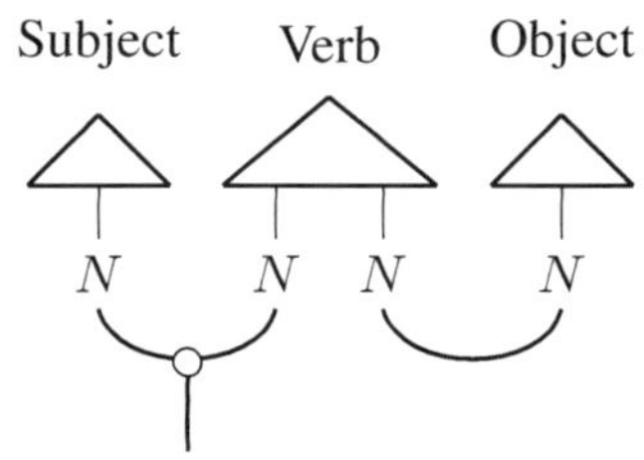

The abstract vectors corresponding to such diagrams are similar to the truth-theoretic case, with the difference that the vectors are populated from corpora and the scalar weights for noun vectors are not necessarily 1 or 0. For subject and object noun context vectors computed from a corpus as follows:

$$\overrightarrow{\text{Subject}} = \sum_k C_k \overrightarrow{n}_k \qquad \overrightarrow{\text{Object}} = \sum_l C_l \overrightarrow{n}_l$$

and the verb a linear map:

$$\overline{\text{Verb}} = \sum_{ij} C_{ij} \overrightarrow{n}_i \otimes \overrightarrow{n}_j$$

computed as above, the concrete meaning of a noun phrase modified by a subject relative clause is as follows:

$$\sum_{kijl} C_k C_{ij} C_l \mu_N(\overrightarrow{n}_k \otimes \overrightarrow{n}_i)\epsilon_N(\overrightarrow{n}_j \otimes \overrightarrow{n}_l)$$
$$= \sum_{kijl} C_k C_{ij} C_l \delta_{ki} \overrightarrow{n}_k \delta_{jl}$$
$$= \sum_{kl} C_k C_{kl} C_l \overrightarrow{n}_k$$

Comparing this to the truth-theoretic case, we see that the previous α_{kl} are now obtained from a corpus and instantiated to $C_k C_{kl} C_l$. To see how the above expression represents the meaning of the noun phrase, decompose it into the following:

$$\sum_k C_k \overrightarrow{n}_k \odot \sum_{kl} C_{kl} C_l \overrightarrow{n}_l$$

Note that the second term of the above, which is the application of the verb to the object, modifies the subject via point-wise multiplication. A similar result arises for the object relative clause case.

As an example, suppose that N has two dimensions with basis vectors $\overrightarrow{n}_1$ and $\overrightarrow{n}_2$, and consider the noun phrase "dog that bites men". Define the vectors of "dog" and "men" as follows:

$$\overrightarrow{\text{dog}} = d_1 \overrightarrow{n}_1 + d_2 \overrightarrow{n}_2 \qquad \overrightarrow{\text{men}} = m_1 \overrightarrow{n}_1 + m_2 \overrightarrow{n}_2$$

and the matrix of "bites" by:

$$b_{11} \overrightarrow{n}_1 \otimes \overrightarrow{n}_2 + b_{12} \overrightarrow{n}_1 \otimes \overrightarrow{n}_2 + b_{21} \overrightarrow{n}_2 \otimes \overrightarrow{n}_1 + b_{22} \overrightarrow{n}_2 \otimes \overrightarrow{n}_2$$

Then the meaning of the noun phrase becomes:

$$\overrightarrow{\text{dog that bites men}} :=$$
$$d_1 b_{11} m_1 \overrightarrow{n}_1 + d_1 b_{12} m_2 \overrightarrow{n}_1 + d_2 b_{21} m_1 \overrightarrow{n}_2$$
$$+ d_2 b_{22} m_2 \overrightarrow{n}_2 = (d_1 \overrightarrow{n}_1 + d_2 \overrightarrow{n}_2) \odot$$
$$((b_{11} m_1 + b_{12} m_2) \overrightarrow{n}_1 + (b_{21} m_1 + b_{22} m_2) \overrightarrow{n}_2)$$

Using matrix notation, we can decompose the second term further, from which the application of the verb to the object becomes apparent:

$$\begin{pmatrix} b_{11} & b_{12} \\ b_{21} & b_{22} \end{pmatrix} \times \begin{pmatrix} m_1 \\ m_2 \end{pmatrix}$$

Hence for the whole clause we obtain:

$$\overrightarrow{\text{dog}} \odot (\overline{\text{bites}} \times \overrightarrow{\text{men}})$$

Again this result is highly intuitive: assuming that the basis vectors of the noun space represent properties of nouns, the meaning of "dog that bites men" is a vector representing the properties of dogs, which have been modified (via multiplication) by those properties of individuals which bite men. Put another way, those properties of dogs which overlap with properties of biting things get accentuated.

7 Conclusion and Future Directions

In this paper, we have extended the compact categorical semantics of Coecke et al. (2010) to analyse meanings of relative clauses in English from a vector space point of view. The resulting vector space semantics of the pronouns and clauses is based on the Frobenius algebraic operations on vector spaces: they reveal the internal structure, or what we call *anatomy*, of the relative clauses.

The methodology pursued in this paper and the Frobenius operations can be used to provide semantics for other relative pronouns and also other closed-class words such as prepositions and determiners. In each case, the grammatical type of the word and a detailed analysis of the role of these words in the meaning of the phrases in which they occur would be needed. In some cases, it may be necessary to introduce a linear map to represent the meaning of the word, for instance to distinguish the preposition *on* from *in*.

The contribution of this paper is best demonstrated via the string diagrammatic representations of the vector space meanings of these clauses. A noun phrase modified by a subject relative clause, which before this paper was depicted as follows:

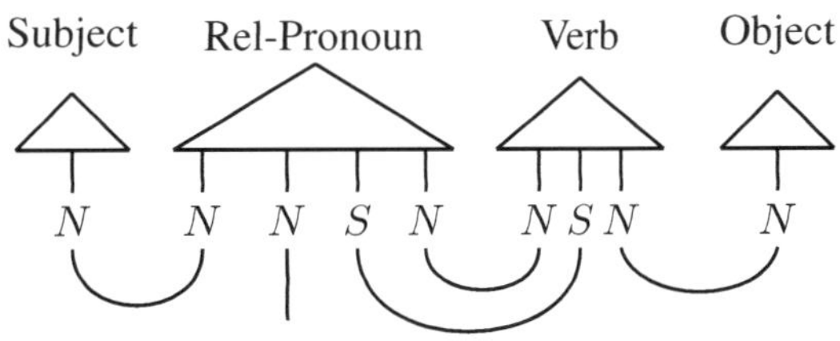

will now include the internal anatomy of its relative pronoun:

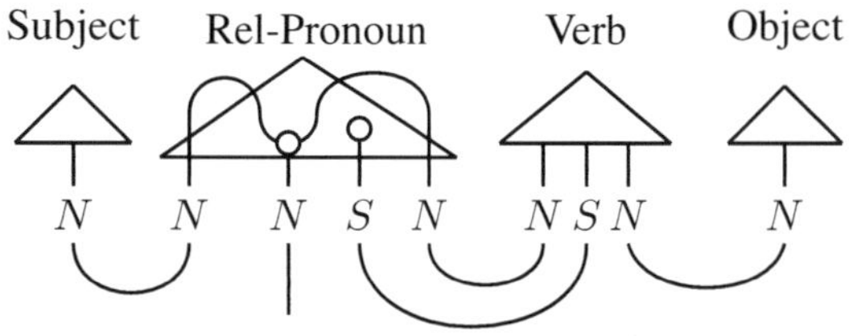

This internal structure shows how the information from the noun flows through the relative pronoun to the rest of the clause and how it interacts with the other words. We have instantiated this vector space semantics using truth-theoretic and corpus-based examples.

One aspect of our example spaces which means that they work particularly well is that the sentence dimension in the verb is already discarded, which means that the ι maps are not required (as discussed above). Another feature is that the simple nature of the models means that the μ map does not lose any information, even though it takes the diagonal of a matrix and hence in general throws information away. The effect of the ι and μ maps in more complex representations of the verb remains to be studied in future work.

On the practical side, what we offer in this paper is a method for building appropriate vector representations for relative clauses. As a result, when presented with a relative clause, we are able to build a vector for it, only by relying on the vector representations of the words in the clause and the grammatical role of the relative pronoun. We do not need to retrieve information from a corpus to be able to build a vector or linear map for the relative pronoun, neither will we end up having to discard the pronoun and ignore the role that it plays in the meaning of the clause (which was perhaps the best option available before this paper). However, the Frobenius approach and our claim that the resulting vectors are 'appropriate' requires an empirical evaluation. Tasks such as the term definition task from Kartsaklis et al. (2013) (which also uses Frobenius algebras but for a different purpose) are an obvious place to start. More generally, the subfield of compositional distributional semantics is a growing and active one (Mitchell and Lapata, 2008; Baroni and Zamparelli, 2010; Zanzotto et al., 2010; Socher et al., 2011), for which we argue that high-level mathematical investigations such as this paper, and also Clarke (2008), can play a crucial role.

Acknowledgements

We would like to thank Dimitri Kartsaklis and Laura Rimell for helpful comments. Stephen Clark was supported by ERC Starting Grant DisCoTex (30692). Bob Coecke and Stephen Clark are supported by EPSRC Grant EP/I037512/1. Mehrnoosh Sadrzadeh is supported by an EPSRC CAF EP/J002607/1.

References

J.C. Baez and J. Dolan. 1995. Higher-dimensional algebra and topological quantum field theory. *Journal of Mathematical Physics*, 36:6073–6105.

M. Baroni and R. Zamparelli. 2010. Nouns are vectors, adjectives are matrices: Representing adjective-noun constructions in semantic space. In *Conference on Empirical Methods in Natural Language Processing (EMNLP-10)*, Cambridge, MA.

A. Carboni and R. F. C. Walters. 1987. Cartesian bicategories. I. *J. Pure and Appied Algebra*, 49:11–32.

S. Clark. 2013. Type-driven syntax and semantics for composing meaning vectors. In Chris Heunen, Mehrnoosh Sadrzadeh, and Edward Grefenstette, editors, *Quantum Physics and Linguistics: A Compositional, Diagrammatic Discourse*, pages 359–377. Oxford University Press.

D. Clarke. 2008. *Context-theoretic Semantics for Natural Language: An Algebraic Framework*. Ph.D. thesis, University of Sussex.

B. Coecke and E. Paquette. 2008. Introducing categories to the practicing physicist. In B. Coecke, editor, *New Structures for Physics*, volume 813 of *Lecture Notes in Physics*, pages 167–271. Springer.

B. Coecke, D. Pavlovic, and J. Vicary. 2008. A new description of orthogonal bases. *Mathematical Structures in Computer Science*, 1:269–272.

B. Coecke, M. Sadrzadeh, and S. Clark. 2010. Mathematical foundations for a compositional distributional model of meaning. *Linguistic Analysis*, 36:345–384.

F. G. Frobenius. 1903. *Theorie der hyperkomplexen Größen*. Preussische Akademie der Wissenschaften Berlin: Sitzungsberichte der Preußischen Akademie der Wissenschaften zu Berlin. Reichsdr.

E. Grefenstette and M. Sadrzadeh. 2011a. Experimental support for a categorical compositional distributional model of meaning. In *Proceedings of Conference on Empirical Methods in Natural Language Processing (EMNLP)*, pages 1394–1404.

E. Grefenstette and M. Sadrzadeh. 2011b. Experimenting with transitive verbs in a discocat. In *Proceedings of the Workshop on Geometrical Models of Natural Language Semantics (GEMS)*.

A. Joyal and R. Street. 1991. The Geometry of Tensor Calculus, I. *Advances in Mathematics*, 88:55–112.

D. Kartsaklis, M. Sadrzadeh, S. Pulman, and B. Coecke. 2013. Reasoning about meaning in natural language with compact closed categories and frobenius algebras. In J. Chubb, A. Eskandarian, and V. Harizanov, editors, *Logic and Algebraic Structures in Quantum Computing and Information*, Association for Symbolic Logic Lecture Notes in Logic. Cambridge University Press.

G. M. Kelly and M. L. Laplaza. 1980. Coherence for compact closed categories. *Journal of Pure and Applied Algebra*, 19:193–213.

J. Kock. 2003. *Frobenius algebras and 2D topological quantum field theories*, volume 59 of *London Mathematical Society student texts*. Cambridge University Press.

J. Lambek. 1958. The Mathematics of Sentence Structure. *American Mathematics Monthly*, 65:154–170.

J. Lambek. 1999. Type Grammar Revisited Logical Aspects of Computational Linguistics. In *Logical Aspects of Computational Linguistics*, volume 1582 of *Lecture Notes in Computer Science*, pages 1–27. Springer Berlin / Heidelberg.

J. Mitchell and M. Lapata. 2008. Vector-based models of semantic composition. In *Proceedings of ACL-08*, pages 236–244, Columbus, OH.

R. Montague. 1974. English as a formal language. In R. H. Thomason, editor, *Formal philosophy: Selected Papers of Richard Montague*, pages 189–223. Yale University Press.

A. Preller and J. Lambek. 2007. Free compact 2-categories. *Mathematical Structures in Computer Science*, 17:309–340.

H. Schütze. 1998. Automatic word sense discrimination. *Computational Linguistics*, 24:97–123.

R. Socher, J. Pennington, E. Huang, A. Y. Ng, and C. D. Manning. 2011. Semi-supervised recursive autoencoders for predicting sentiment distributions. In *Proceedings of the Conference on Empirical Methods in Natural Language Processing*, Edinburgh, UK.

F. M. Zanzotto, I. Korkontzelos, F. Fallucchi, and S. Manandhar. 2010. Estimating linear models for compositional distributional semantics. In *Proceedings of COLING*, Beijing, China.

Vowel Harmony and Subsequentiality

Jeffrey Heinz and **Regine Lai**
University of Delaware
{heinz,rlai}@udel.edu

Abstract

Attested and 'pathological' vowel harmony patterns are studied in the context of subclasses of regular functions. The analysis suggests that the computational complexity of phonology can be reduced from regular to weakly deterministic.

1 Introduction

The expressivity of ordered rewrite-rule grammars for phonology (Chomsky and Halle, 1968, henceforth SPE) and two-level phonology (Koskenniemi, 1983) are exactly the class of regular relations (Johnson, 1972; Kaplan and Kay, 1994; Beesley and Kartunnen, 2003). Since SPE-style grammars can express virtually any phonological generalization, it follows that the generalizations themselves are regular, even if they are represented with other formalisms (such as OT grammars).

This result can be interpreted cognitively as establishing a universal property of phonological patterns: humanly possible phonological patterns are regular. If correct, this would mean, for example, that humanly possible syntactic patterns which are nonregular are not humanly possible *phonological* patterns (Heinz and Idsardi, 2011; Lai, 2012; Heinz and Idsardi, 2013).

Recent research suggests that stronger universals than "being regular" can be established for phonology. It has been shown that segmental phonotactic patterns are star-free (Heinz et al., 2011), as are virtually all stress patterns (Rogers et al., to appear), and the semantics of two-level rules appear to ensure that these mappings have star-free-like properties, provided the contexts to the rules are star-free (Yli-Jyrä and Koskenniemi, 2006).[1]

This paper examines the hypothesis that subsequentiality is a necessary property of phonological patterns by studying theories of iterative vowel harmony. Informally, a function is left (right) subsequential if there is a finite-state transducer describing the function which processes strings from left to right (right to left) deterministically on the input. We use the term 'subsequentiality' to mean *either* left or right subsequential.

Previous work has found that synchronically attested metathesis patterns and partial reduplication patterns are either left or right subsequential (Chandlee et al., 2012; Chandlee and Heinz, 2012). Also Gainor et al. (2012) show that the vowel harmony generalizations in Nevins (2010) are also left or right subsequential mappings. Gainor et al.'s analysis, while insightful, is incomplete since Nevin's theory of vowel harmony invokes underspecification and other theories of vowel harmony do not. Phonological underspecification is explained in section 3. The linguistic generalizations examined in this paper come from two types of theories of vowel harmony patterns in linguistics which do not use underspecification: traditional directional theories and dominant/recessive/stem-control theories.

We prove that subsequentiality separates directional theories from logically possible but 'pathological' vowel harmony patterns (Wilson 2003). (This claim was also made by Gainor et al. without proof.) It is also shown that dominant/recessive/stem-control theories posit generalizations which are neither left nor right subsequential, but which are *weakly deterministic*. Informally, this means that these generalizations can be decomposed into a left subsequential and right subsequential function without the left-subsequential function marking up its output in any special way. We conjecture this is

[1]There are multiple ways to generalize the class of star-free regular sets to regular relations (Benedikt et al., 2001) so it would be interesting to determine more exactly the nature of such two-level rules.

not the case for the pathological patterns. Since subsequential and weakly deterministic functions are proper subclasses of regular relations, these results suggest concretely how the computational complexity of phonology established by earlier researchers can be improved.

Mathematical and phonological preliminaries are given in sections 2 and 3, respectively. Sections 4, 5, and 6 consider the vowel harmony patterns with respect to the regular and subsequential boundaries. and weakly deterministic boundaries, respectively. Section 7 concludes.

2 Preliminaries

2.1 Regular relations and functions

If X denotes a finite alphabet then X^* and X^n denotes the sets of all finite strings and the set of all strings of length n over X, respectively. The length of a string w is $|w|$. The unique string of length zero is denoted λ. A string w of length k can be written $w_1 w_2 \cdots w_k$, where w_i is the ith letter of w. The reverse of a string $w = w_1 \cdots w_k$ is $w^r = w_k \cdots w_1$. For finite alphabets X and Y, a *relation* is a subset of $X^* \times Y^*$. If R is a relation, the *reverse relation* $R^r = \{(x^r, y^r) \mid (x,y) \in R\}$. Note the reverse relation is *not* the inverse relation. A relation R is *length-preserving* iff for all $(x,y) \in R$ it is the case that $|x| = |y|$. It is *length-increasing* iff there exists $(x,y) \in R$ such that $|x| < |y|$.

Any relation $R \subseteq X^* \times Y^*$ is a *function* iff for all $x \in X^*$, there is at most $y \in Y^*$ such that $(x,y) \in R$. In this case, we often write $R : X^* \to Y^*$. For two functions $f : X^* \to Y^*$ and $g : Y^* \to Z^*$, the *composition* of f and g is a function $h : X^* \to Z^*$ such that $h(x) = g(f(x))$. We write $h = g \circ f$.

For all $x \in X^*$, the prefixes of x are $Pr(x) = \{u \in X^* \mid \exists v \in X^* \text{ such that } x = uv\}$. For any set $L \subseteq X^*$, the longest common prefix of L is

$$lcp(L) = w \Leftrightarrow w \in \bigcap_{x \in L} Pr(x) \quad \wedge \\ (\forall w' \in \bigcap_{x \in L} Pr(x))\big[|w'| \leq |w|\big] \tag{1}$$

Regular relations are those describable by *finite-state transducers* (FSTs).[2] A finite-state transducer T is a tuple (Q, X, Y, I, F, δ) where Q is a finite set of states, X and Y are finite alphabets, $I, F \subseteq Q$ are the initial and final states, respectively, and $\delta \subseteq Q \times X^* \times Y^* \times Q$ is the transition function. For all FSTs, the transition function δ is recursively extended to δ^* in the usual way. The relation that a finite state transducer $T = (Q, X, Y, I, F, \delta)$ recognizes/accepts/-generates is

$$R(T) = \Big\{ (x,y) \in X^* \times Y^* \mid (\exists q_i \in I) \\ (q_f \in F)\big[(q_i, x, y, q_f) \in \delta^*\big]\Big\}. \tag{2}$$

Let RR denote the class of regular relations. This class is closed under concatenation, Kleene closure, union, composition, and inversion, but not intersection or complement.

A *subsequential* transducer is a tuple $(Q, q_0, X, Y, \sigma, \delta)$, where Q is a finite set of states, X and Y are finite alphabets, $q_0 \in Q$ is the initial state, and $\sigma \subseteq Q \times Y^*$ is the output function. Informally, subsequential transducers are weighted acceptors that are deterministic on the input, and where the weights are strings and multiplication is concatenation.

Formally, the transition function $\delta \subseteq Q \times X \times Y^* \times Q$ is deterministic:

$$(q, a, u, r), (q, a, v, s) \in \delta \Rightarrow u = v \wedge r = s.$$

The transition function δ is also recursively extended to δ^*. The relation that a subsequential transducer $T = (Q, q_0, X, Y, \sigma, \delta)$ recognizes/accepts/generates is

$$R(t) = \Big\{ (x, yz) \in X^* \times Y^* \mid (\exists q \in F) \\ \big[(q_0, x, y, q) \in \delta^* \wedge z = \sigma(q)\big]\Big\}. \tag{3}$$

Since subsequential transducers are deterministic, the relations they recognize are functions.

Functions recognized by subsequential transducers are called *left subsequential*. A function f is *right subsequential* iff its reverse f^r is left subsequential. Observe that for all $w \in X^*$, the image of a right subsequential function f of w can be calculated by reversing w, processing the result with the subsequential transducer T recognizing f^r, and then reversing the result. Formally,

$$(\forall w \in X^*)[f(w) = T(w^r)^r]. \tag{4}$$

However, there is another way to state the above without the reversing function $(\cdot^r)$. This is to recognize that $f(w)$ can be computed by applying T

[2]In the algebraic theory of automata, these are called *rational* relations (Berstel, 1979; Sakarovitch, 2009).

to w from right to left, instead of from left to right. When T processes w right to left, we write $\overleftarrow{T}(w)$ and when T processes w left to right, we write $\overrightarrow{T}(w)$. Then we can restate Equation 4 as

$$(\forall w \in X^*)[f(w) = \overleftarrow{T}(w)]. \quad (5)$$

Let LSF and RSF denote the class of left and right subsequential functions, respectively.

Theorem 1 (Mohri 1997) *The following hold:*

1. $LSF, RSF \subsetneq RR$.
2. $RSF^r = LSF$.
3. *LSF and RSF are incomparable.*

There are some relevant subclasses and generalizations of subsequentiality. A subsequential transduction T is *sequential* iff for all $q \in Q$, it is the case that $\sigma(q) = \lambda$.[3] Mohri (1997) generalizes subsequentiality to p-subsequentiality (allowing up to p outputs for each input), preserving many important properties. Mohri's generalizations are important here because there are likely to be a bounded number of exceptions, or optional forms, in actual vowel harmony systems that fall outside the purview of the 1-subsequential analysis presented here, but which would presumably not fall outside a p-subsequential analysis (not presented here).

Elgot and Mezei proved the following.

Theorem 2 (Elgot and Mezei 1965) *Let $T : X^* \to Y^*$ be a function. Then $T \in RR$ iff there exists $L : X^* \to Z^* \in LSF$, and $R : Z^* \to Y^* \in RSF$ with $X \subseteq Z$ such that $T = R \circ L$.*

What this decomposition means is that the computation of $T(x) = y$ can be accomplished by (1) reading x sequentially from left to right with a subsequential transducer, which transforms it into a word z possibly marking it up with additional symbols; (2) reading the resulting word z from right to left with another subsequential transducer and writing from right to left the final output y. As their proof makes clear, this decomposition of T is possible because the alphabet Z may be strictly larger than X, and so z can be marked-up with additional symbols which carry additional information.[4]

Finally, we review one important property of subsequential transducers and regular sets. For any function $f : X^* \to Y^*$ and $x \in X^*$, let the *tails* of x in f be defined as

$$TL_f(x) = \{(y, v) \mid f(xy) = uv \wedge u = lcp(f(xX^*))\} . \quad (6)$$

Every subsequential transducer T computing a function f admits a *canonical* form, where the states of T are in one-to-one correspondence with $TL_f(x)$ for all $x \in X^*$.

Theorem 3 (Oncina et al. 1993) $f \in LSF \Leftrightarrow \{TL_f(x) \mid x \in X^*\}$ *has finite cardinality.*

This theorem is the functional counterpart to the Myhill/Nerode relation. Recall that for any set of strings L, the tails of a word w with respect to L is defined as $TL_L(w) = \{u \mid wu \in L\}$. This relation partitions the set of all logically possible strings into a finite set of equivalence classes iff the set L is regular. These equivalence classes are the basis for constructing the smallest deterministic acceptor for a regular language.

Similarly, in the construction of the canonical subsequential transducer for a left subsequential function, the states correspond to the sets of tails defined in (6) above. There is a rich literature on subsequential functions (Elgot and Mezei, 1965; Berstel, 1979; Oncina et al., 1993; Mohri, 1997; Roche and Schabes, ; Sakarovitch, 2009).

2.2 Weak Determinism

Here we introduce the notion of *weak determinism*. Informally, these are regular functions which decompose into left and right subsequential functions as in Elgot and Mezei's theorem but *without* the mark-up given by the intermediate, larger alphabet. Thus, they are not necessarily deterministic, but they are "more" deterministic than regular functions where Elgot and Mezei decomposition *requires* the intermediate mark-up.

While the mark-up can be accomplished by introducing new symbols (as done in Elgot and

[3]Sakarovitch (2009) prefers the term 'sequential' for subsequential functions and the term 'pure sequential' for sequential functions. While his arguments are reasonable (pp. 651-2), we adopt the more widely adopted terminology.

[4]Berstel (1979) provides an updated proof. This book is out of print but the first four chapters are available for download at `http://www-igm.univ-mlv.fr/~berstel/LivreTransductions/LivreTransductions.html`. The theorem and proof begin on page 117 in the online version and on page 126 in the printed version.

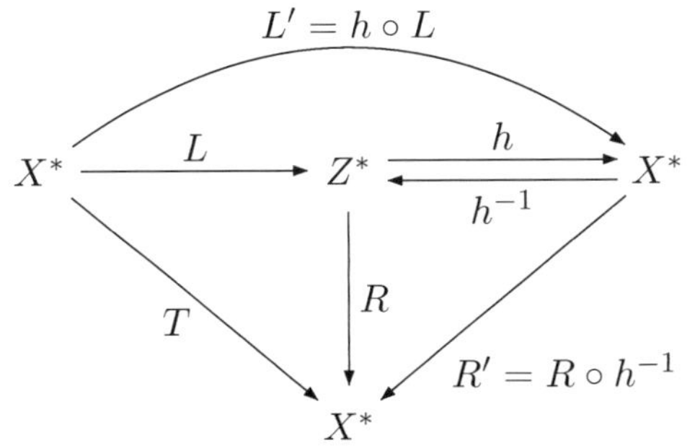

Figure 1: Decompositions of regular function T with $X \subseteq Z$.

Mezei's proof), for any alphabet with at least two symbols, the mark-up can also be accomplished by *coding* these new symbols as strings formed over the original alphabet.[5] Figure 1 illustrates. Let X be a finite alphabet containing the symbols $\{a, b\}$. Consider any *alphabet-preserving* regular function $T : X^* \to X^*$. By Elgot and Mezei's theorem, there exists left subsequential $L : X^* \to Z^*$ and right subsequential $R : Z^* \to X^*$ with $X \subseteq Z$ such that $T = R \circ L$. Let $h : Z^* \to X^*$ be a function which encodes each word w in Z^* by coding each symbol in w as follows. Assume some enumeration of the symbols in Z and the rewrite the nth symbol of Z as ab^na. For example if $Z = \{a, b, c\}$ and $w = cab$ then $h(w) = abbaaaaba$. It is not difficult to verify that h is length-increasing and that both h and h^{-1} are subsequential functions. Letting $R' = R \circ h^{-1}$ and $L' = h \circ L$, it follows that $T = R' \circ L'$ and that both R' and L' have domain and co-domain X^*.

For this reason, it is not sufficient to require that the decomposition be *alphabet-preserving* (i.e. $Z = X$) to avoid any mark-up. It is also necessary that the first factor L not be *length-increasing*. This is because the only way to unambiguously encode a larger alphabet into a smaller one is with length-increasing functions (like h in the above example).

Definition 1 *A regular function T is* weakly deterministic *iff there exists $L : X^* \to X^* \in LSF$, and $R : X^* \to X^* \in RSF$ such that L is not length-increasing and $T = R \circ L$. The class of weakly deterministic functions is denoted* WD.

The corollary below is immediate from this definition and Elgot and Mezei's theorem.

Corollary 1 $LSF, RSF \subseteq WD \subseteq RR$.

[5]We are indebted to an anonymous reviewer for this important observation.

	noun	genitive	gloss
a.	ip	ip-in	*rope*
b.	el	el-in	*and*
c.	son	son-un	*end*
d.	pul	pul-un	*stamp*

Table 1: Examples illustrating a fragment of the Vowel harmony from Turkish (Nevins 2010:32).

(a)

w	$f(w)$
/ip-un/	[ip-in]
/el-un/	[el-in]
/son-un/	[son-un]
/pul-un/	[pul-un]
...	

(b)

w	$f(w)$
/−C+C/	[−C−C]
/C+C+C/	[C+C+C]
...	

Table 2: Examples showing fragments of the phonological function describing Turkish back harmony assuming the underlying genitive morpheme is /-un/.

The vowel harmony analysis below is sufficient to go a step further and demonstrate a separation between LSF, RSF on one side and WD on the other. We conjecture that one unattested 'pathological' vowel harmony patterns separates WD from RR.

3 Vowel Harmony

Vowel harmony is a pattern wherein vowels assimilate with respect to some feature. Table 1 shows two allomorphs of the genitive suffix, [-in] and [-un]. The allomorph is predictable based on the front/back dimension of the preceding vowel: if the preceding vowel is front then [-in] occurs, but if it is back then [-un] occurs. (Turkish also has rounding harmony, which is not shown here.)

Phonological analysis conceives of the phonological grammar as a function which maps an abstract lexical representation (the 'underlying form') to a more concrete—but still abstract—phonological representation (the 'surface form') (Hyman, 1975; Kenstowicz, 1994; Hayes, 2009). Phonological transcriptions, like the ones in Table 1, represent surface forms.

To illustrate, consider a simple phonological analysis of the Turkish forms above, which posits the underlying form of the genitive suffix to be /-un/ and a mapping f which derives the surface forms as shown in Table 2(a). Table 2(b) describes the mapping only in terms of the relevant details where [+] indicates [+ back] vowels, [−] indicates [$-$back] vowels, and C consonants.

Nevins' (2010) analysis of vowel harmony utilizes underspecification. We illustrate this concept ostensively with the Turkish example above. Instead of positing the underlying form of the suffix to be either /-in/ or /-un/, underspecification theory would posit it to be /-Vn/ where V is high, unrounded vowel unspecified for backness. In Nevins' theory, underlying vowels which have feature specifications can spread those features only to vowels unspecified for those features. Underspecification is not congruent with research in Optimality Theory (Prince and Smolensky, 2004), which, by the principle of the rich base, requires every underlying form to be considered (including those where every vowel is fully specified). Gainor et al. (2012) show that the iterative mappings Nevins describes are subsequential, but this does not address those theories (like OT) which may not consider underlying forms to permit underspecification. All the vowel harmony patterns considered in this paper do not admit any underspecification whatsoever.

A traditional view of vowel harmony analyzes vowel harmony patterns as either instances of *progressive harmony* (PH) or *regressive harmony* (RH). Informally, progressive harmony means the value of a feature can be thought to spread from left to the right (as in the Turkish example above). Conversely, regressive harmony can be thought of as spreading from right to left. This is illustrated with examples (a-d) in Table 3.

Other theories of vowel harmony reject that directionality is a primitive of the theory and argue that vowel harmony is either *dominant/recessive* (DR) or *stem-controlled* (SC) (Baković, 2000; Krämer, 2003) (see also (Archangeli and Pulleyblank, 1994)). Dominant/recessive theories analyze vowel harmony patterns by postulating that a particular feature value of a harmonizing feature is the dominant one. The DR function in Table 3 identifies the $[+]$ value as the dominant one; so any underlying representation containing the harmonizing feature with the value $[+]$ will surface so that the harmonizing feature in all vowels will also be $[+]$. Stem-controlled analyses are similar to dominant-recessive theories, however the feature that spreads is determined not by its value but instead by the morphological position of the vowel to which the feature belongs (for instance, is the vowel in a stem or affix?).

An additional complication is that variations of the above functions are introduced by *neutral* vowels, which never undergo harmony. They come in two kinds: *transparent* vowels which permit features to spread through them, and *opaque* vowels, which block the spread of harmony, but trigger their own harmony domain. Some effects of neutral vowels are shown in rows (e-f) in Table 3. (Symbols [$\boxminus$] and [$\ominus$] are [$-$F] vowels that are opaque and transparent, respectively. Likewise we use [$\boxplus$] and [$\oplus$] to denote opaque and transparent vowels which are [$+$F].)

Additionally, the phonological literature includes discussion of logically possible, unattested and unnatural vowel harmony patterns that are predicted by classical approaches to vowel harmony in OT. These patterns include *sour grapes* (SG) (Padgett, 1995; Wilson, 2003) and *majority rules* (MR) (Lombardi, 1999; Baković, 2000). Informally, SG is like progressive harmony except that vowels only harmonize if every vowel is guaranteed to harmonize. For example, if an opaque vowel occurs after the trigger, no non-neutral vowel harmonizes with the trigger. Majority Rules instantiates the following rule: If the number of segments with αF is greater than the number of segments with $-\alpha F$, then segments with αF are the triggers of harmony and segments with $-\alpha F$ are the targets and undergo change. Because phonologists consider SG and MR to be bizarre, they are referred to as 'pathologies' (Wilson, 2003; Wilson, 2004; Finley, 2008) and it is a strike against a theory if it predicts the existence of either SG or MR.

Henceforth, let $X = Y = \{+, -, C, \boxminus, \boxplus\}$. These symbols represent equivalence classes of a partition of the phonemic inventory of any language which exhibits progressive harmony for the feature F. The symbols $+$ and $-$ represent the classes of all harmonizing vowels which are $+F$ and $-F$, respectively. Phonemes invisible to harmony are in the C class; this includes consonants and transparent vowels. The symbol $\boxplus$ ($\boxminus$) refers to opaque vowels which are $+F$ ($-F$), which block the spread of $-F$ ($+F$), and which begins a new harmonic domain spreading $+F$ ($-F$).

The vowel harmony mappings in this paper are all, in fact, same-length relations. Furthermore, they are sequential. These additional properties do not appear to be shared by other phonological processes. Epenthesis and deletion are common

	w	PH(w)	RH(w)	DR(w)	SG(w)	MR(w)
a.	$/+--/$	$[+++]$	$[---]$	$[+++]$	$[+++]$	$[---]$
b.	$/-++/$	$[---]$	$[+++]$	$[+++]$	$[---]$	$[+++]$
c.	$/---/$	$[---]$	$[---]$	$[---]$	$[---]$	$[---]$
d.	$/-+-/$	$[---]$	$[---]$	$[+++]$	$[---]$	$[---]$
e.	$/+-\boxminus/$	$[++\boxminus]$	$[--\boxminus]$	$[++\boxminus]$	$[+-\boxminus]$	$[--\boxminus]$
f.	$/+\ominus-/$	$[+\ominus+]$	$[-\ominus-]$	$[+\ominus+]$	$[+\ominus+]$	$[-\ominus-]$

Table 3: Example mappings of underlying forms (w) given by progressive harmony (PH), regressive harmony (RH), dominant/recessive harmony (DR), sour grapes harmony (SG), and majority rules harmony (MR). Symbols [+] indicates a [+F] vowel and [−] indicates a [−F] vowel where "F" is the feature harmonizing. Symbols [$\boxminus$] and [$\ominus$] are [−F] vowels that are opaque and transparent, respectively.

cross-linguistically, and the metathesis patterns analyzed by Chandlee and Heinz (2012) are not sequential (though they are subsequential). For this reason, we keep the analysis focused at the level of subsequentiality.

4 The regular boundary

In this section we show that the regular boundary is sufficient to distinguish the pathological Majority Rules pattern from the attested progressive and regressive harmony patterns.

Formally, MR functions can be defined as follows. Let $|w|_{+F}$ and $|w|_{-F}$ denote the number of participating vowels (i.e. non transparent vowels) which are $+F$ and $-F$, respectively, in $w \in X^*$. Then we define a Majority Rules Harmony pattern as any same-length function which at a minimum obeys the following:

$$MR(w) = \begin{cases} +^{|w|} \text{ if } |w|_{+F} > |w|_{-F} \\ -^{|w|} \text{ if } |w|_{-F} > |w|_{+F} \end{cases} \quad (7)$$

The result below seems to be widely known (see Riggle (2004, chapter 7, section 5)) though we have not been able to find a proof in print.

Theorem 4 *Majority Rules is not regular.*

Proof By way of contradiction, suppose that MR is a regular relation. Since regular relations are closed under inverse, so is MR^{-1}. The image of a regular set under a regular relation is also a regular set (see Roche and Schabes (, pp. 41-43)). Therefore, $MR^{-1}(+^*)$ is a regular set. Since regular sets are closed under intersection, it follows that $MR^{-1}(+^*) \cap (+^*-^*)$ is regular as well. Call this set S.

However, S is in fact *not* a regular set. Since MR is length preserving, for all odd $k \in \mathbb{N}$, it is the case that $MR(+^k) \cap (+^*-^*) = +^m-^n$ where $0 \leq m, n \leq k$ and $m + n = k$ and $m > n$. Furthermore, for all for all odd k, it is the case that $TL_S(+^k)$, includes $-^n$ for all $n < k - 1$ but excludes $-^n$ for all $n > k + 1$. Thus there is a distinct Nerode-equivalence class for each odd k, and hence S is not a regular set, and therefore MR is not a regular relation. □

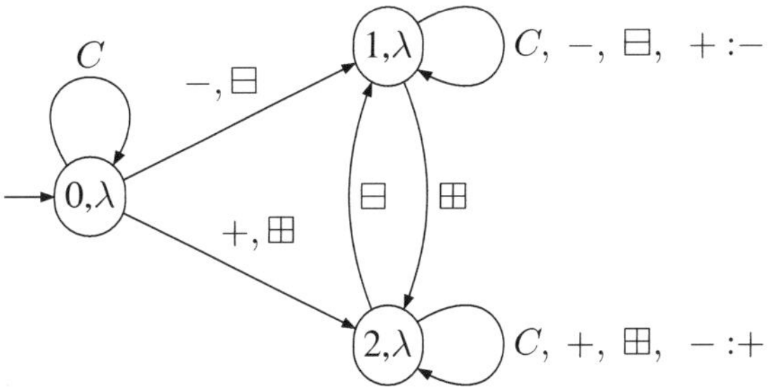

Figure 2: A subsequential transducer which recognizes iterative, progressive harmony.

On the other hand, progressive and regressive harmony are regular; in fact, subsequential. For concreteness we analyze a canonical progressive harmony pattern which includes neutral vowels.

The subsequential transducer T_{PH} in Figure 2 faithfully captures the PH function. Labels on the transitions are interpreted as in Beesley and Karttunen (2003): commas delimit multiple transitions; the label x:y means x is the input and y the output; absence of a colon means the input and output are identical. The rightmost symbol interior to a state is the output of the σ function.

While T_{PH} presupposes languages have opaque vowels, it can be modified as needed to remove this assumption without losing subsequentiality. If a language has no opaque vowels, those transitions can be removed. Since any subgraph of the transducer shown in Figure 2 is also subsequential, this establishes the left subsequentiality of iterative, progressive harmony patterns without underspecification.[6]

As for iterative regressive harmony patterns, they are simply the reverse of iterative progressive patterns. In other words, for all $w \in \Sigma^*$, $RH(w) = \overleftarrow{T_{PH}}$.

5 The subsequential boundary

In this section we show that while the regular boundary is not sufficient to separate the pathological Sour Grapes pattern from attested harmony patterns, the subsequential boundary *is* sufficient.

Padgett (1995) defines *Sour Grapes Harmony* as "Either all features must spread, or none will..." For concreteness, consider a progressive Sour Grapes pattern. The form $+ - --$ would be mapped to $+ + ++$ as in progressive harmony, but the form $+ - -\boxminus$ is mapped to $+ - -\boxminus$ because the opaque vowel will not become $+F$, and so the spreading process grumpily chooses not to spread at all. Therefore, a progressive Sour Grapes Harmony pattern is defined as any length-preserving function which at a minimum includes the following mappings for all $n \in \mathbb{N}$:

$$SG(+-^n) = ++^n \wedge SG(+-^n\boxminus) = +-^n\boxminus \quad (8)$$

There is a finite state transducer which describes this fragment of the SG function, shown in Figure 3. As a total function, for SG to be regular, it is important that the image of the complement of $\cup_{n\in\mathbb{N}}\{+-^n\} \cup_{n\in\mathbb{N}} \{+ -^n \boxminus\}$ under SG also be regular. Pictorially, this would mean that the fragment shown in Figure 3 is a *subgraph* of the full SG pattern. Crucially, however, there can also be no transition from state 2 bearing the input symbol $\boxminus$ that can lead (even eventually) to a final state.

We now prove the main theorem of this paper.

Theorem 5 *SG is neither left nor right subsequential.*

Proof We show that, for all distinct $n, m \in \mathbb{N}$, the tails of $+-^n$ is not the same as the tails of $+-^m$. This immediately implies that the canonical left subsequential transducer would have infinitely many states, and hence that any SG pattern meeting Equation 8 is not left subsequential.

[6]It is true that T_{PH} does not model progressive harmony patterns where transparent vowels can trigger harmony. It is not difficult to modify T_{PH} to accommodate this without sacrificing subsequentiality.

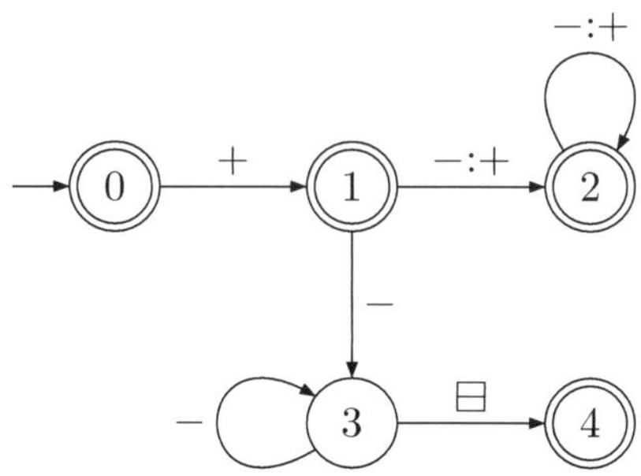

Figure 3: A non-deterministic transducer which recognizes a fragment of SG harmony.

To illustrate, consider $x = +-$. Since $SG(+ - X^*)$ includes elements $+++$ and $+-\boxminus$ (mapped from e.g. $+ - -$ and $+ - \boxminus$, resp.), it follows that $lcp(SG(+ - X^*)) = +$. Therefore, $(-, ++) \in TL_{SG}(+-)$. Observe that $(-, +^n) \notin TL_{SG}(+-)$ for all $n \neq 2$ since SG is length-preserving.

More generally it is the case that $(-, +^{n+1}) \in TL_{SG}(+-^n)$ and $(-, +^m) \notin TL_{SG}(+-^n)$ for all $m \neq n + 1$. Therefore there are infinitely many distinct sets of tails for functions conforming to (8), and thus no SG pattern is subsequential.

A similar argument (omitted) establishes that any SG pattern is not right subsequential. □

Consequently, the subsequential boundary separates SG and MR from PH and RH.

6 The weakly deterministic boundary

As mentioned earlier, the dominant/recessive and stem-control theories of vowel harmony reject the directionality generalizations inherent in the PH and RH mappings. If these theories are correct, then it is important to see what boundary (if any) separates MR and SG from vowel harmony patterns described with dominant/recessive and stem-control based generalizations. We show that the mappings these theories posit are, like SG, not subsequential. However, we believe there is an interesting complexity difference between SG on the one hand and DR and SC on the other. In particular, we show that DR and SC are weakly deterministic. We conjecture that SG is not weakly deterministic and provide the intuition behind this conjecture, though its proof currently escapes us.

A dominant/recessive analysis of vowel harmony says if the word contains the dominant value of the harmonizing feature, then other vowels in the word take on the dominant value for this feature. For example, if the [+] value for harmonizing feature F is the dominant one and an underlying representation contains a vowel specified as +F, then all other non-neutral vowels in the word will be realized as +F as well.

Therefore, we can define a function as dominant/recessive as any length-preserving function which includes the following mappings:

$\forall w \in \{+,-\}^*$,

$$DR(w) = \begin{cases} +^{|w|} & \text{if } (\exists 0 \leq i \leq |w|)[w_i = +] \\ -^{|w|} & \text{otherwise} \end{cases} \tag{9}$$

The next two theorems establish that DR harmony is properly weakly deterministic.

Theorem 6 *DR is neither left nor right subsequential.*

Proof The proof is similar to the one for SG. We show that, for all distinct $n, m \in \mathbb{N}$, the tails of $-^n$ is not the same as the tails of $-^m$.

Consider first $x = --$. To find its tails we must know $lcp(DR(- - X^*))$. Since $DR(- - X^*)$ includes elements $- - -$ and $+ + +$ (mapped from e.g. $- - -$ and $- - +$, resp.), it follows that $lcp(DR(+ - X^*)) = \lambda$. Therefore, $(-, - - -) \in TL_{DR}(--)$. Observe that for all $n \neq 2$, it is the case that $(-, +^n) \notin TL_{DR}(--)$ since DR is length-preserving.

Next consider $x = - - -$. To find its tails we must know $lcp(DR(- - -X^*))$. $DR(- - -X^*)$ includes elements $- - --$ and $+ + ++$ (mapped from e.g. $- - --$ and $- - -+$, resp.), and so again the longest common prefix is λ. Therefore, $(-, + + ++)$ belongs to $TL_{DR}(- - -)$ and $(-, +^n) \notin TL_{DR}(+-)$ for all $n \neq 3$ since DR preserves string lengths.

More generally for all distinct $n, m \in \mathbb{N}$ it is the case that $(-, +^{n+1}) \in TL_{DR}(-^n)$ and $(-, +^m) \notin TL_{DR}(+-^n)$ for all $m \neq n + 1$. Therefore there are infinitely many distinct sets of tails for functions conforming to (9), and thus no DR pattern is subsequential.

A similar argument (omitted) establishes that any DR pattern is not right subsequential. □

On the other hand, DR is weakly deterministic. We establish this for the case when the alphabet contains only $\{+, -\}$. In fact, DR is simply the composition of progressive harmony and regressive harmony where only the dominant feature value spreads. Figure 4 shows a subsequential transducer T_{PHP} describing a progressive harmony function where only $[+]$ spreads. Observe

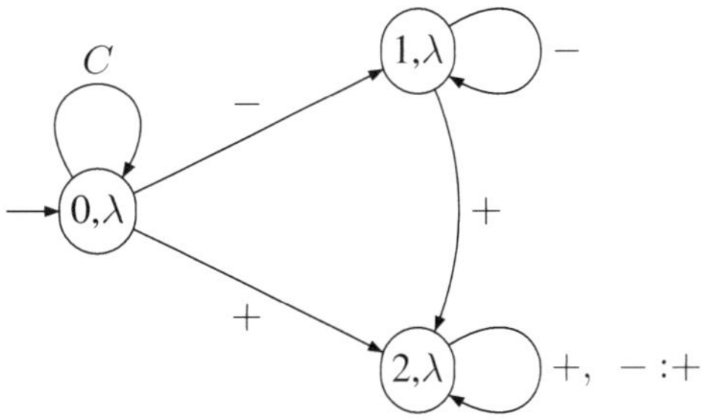

Figure 4: The subsequential transducer T_{PHP} which recognizes iterative, progressive harmony where only the + value spreads.

that the transducer in Figure 4 is nearly identical to T_{PH} in Figure 2 without the opaque vowels and the C symbol. The important difference is that in Figure 2 there is a transition from state 1 to itself which reads a $[+]$ input and outputs a $[-]$, but in Figure 4, the transducer in state 1, upon reading input $[+]$ transitions to state 2 and writes out $[+]$.

Let PHP denote the function T_{PHP} computes.

Theorem 7 *DR is weakly deterministic.*

Proof We show that for all $w \in \{+,-\}^*$, $DR(w) = \overleftarrow{T_{PHP}} \circ \overrightarrow{T_{PHP}}(w)$.

Case 1. There exists $0 \leq i \leq |w|$ such that $w_i = +$. Then $DR(w) = +^{|w|}$. It follows from the definition of PHP that for all $j \geq i$, $w_j = +$. Letting $u = PHP(w)^r$, it follows that $u_1 = +$. Thus $PHP(u) = +^{|w|}$. The reverse of $+^{|w|}$ is clearly itself. Therefore, $PHP^r \circ PHP(w) = +^{|w|}$.

Case 2. Case 1 does not hold. Then $DR(w) = -^{|w|}$. By definition, $PHP(w) = -^{|w|}$. Clearly then $PHP^r \circ PHP(w) = -^{|w|}$.

Since $DR = \overleftarrow{T_{PHP}} \circ \overrightarrow{T_{PHP}}$, since $\overrightarrow{T_{PHP}}$ and $\overleftarrow{T_{PHP}}$ are alphabet-preserving, and since $\overrightarrow{T_{PHP}}$ is not length-increasing, the theorem is proved. □

While, the proof of theorem 7 is limited to words in $\{+,-\}^*$, we believe the extension to words in $\{+, -, C, \boxminus, \boxplus\}^*$ is only challenging technically, and not conceptually. For example, the definition of DR harmony above in Equation 9 should be more articulated so that, for instance, $DR(--\boxminus- -+--\boxminus--) = --\boxminus+++++\boxminus--$.

Stem-controlled analyses are similar to dominant/recessive theories. Unlike dominant/recessive theories, however, the vowels which trigger harmony are the ones which belong to the morphological stem. Table 4 illustrates with stem boundaries indicated with #. What

	w	SC(w)
a.	$+\# - \#-$	$---$
b.	$-\# + \#+$	$+++$
c.	$-\# - \#-$	$---$
d.	$-\# + \#-$	$+++$

Table 4: Example mappings of underlying forms (w) with three vowels given stem control theories of vowel harmony. In each example, the middle vowel is the only vowel in the stem.

happens when there is more than one stem vowel? The stem precedence generalization (Baković, 2003) states that "an alternating affix vowel always agrees with the adjacent vowel in the stem to which the affix is attached." Therefore stem vowels themselves are not targets of the harmony process. Consequently, underlying $/--\#+-+-\#++/$ would surface as $[++\#+-+-\#--]$.

Since every underlying form is assumed to contain a stem, the domain of SC harmony is $X^*\#X^+\#X^*$. Then $\forall w\#u\#v \in X^*\#X^+\#X^*$, it is the case that

$$SC(w) = \begin{cases} +^{|w|}\#u\# +^{|v|} \text{ if } u_1 = + \wedge u_{|u|} = + \\ +^{|w|}\#u\# -^{|v|} \text{ if } u_1 = + \wedge u_{|u|} = - \\ -^{|w|}\#u\# +^{|v|} \text{ if } u_1 = - \wedge u_{|u|} = + \\ -^{|w|}\#u\# -^{|v|} \text{ if } u_1 = - \wedge u_{|u|} = - \end{cases} \quad (10)$$

The analysis of SC is the same as DR; and so the proofs are omitted.

Theorem 8 *SC is neither left nor right subsequential.*

Theorem 9 *SC is weakly deterministic.*

We believe there is a difference between DR and SG: we conjecture that SG is not weakly deterministic. To get an intuition why, consider the two subsequential transducers A and B in Figure 5. Limiting our attention to the domain $\{+-^n\} \cup \{+-^n \boxminus\} \cup \{\lambda\}$, the composition $\overleftarrow{B} \circ \overrightarrow{A}$ equals SG. This is possible because these functions make use of an additional symbol $[\overset{?}{-}]$, indicating a minus value whose left context matches the environment to become $[+]$.

Table 5 illustrates the role the additional symbol plays in the derivation. We are doubtful that it is possible to decompose SG into a left and right deterministic function where the left function is prohibited from marking up its output in any way (either with extra symbols or with a length-increasing coding trick).

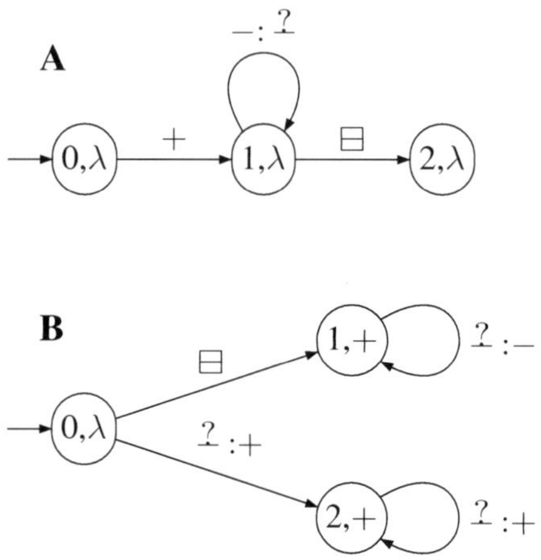

Figure 5: Two subsequential transducers such that $\overleftarrow{B} \circ \overrightarrow{A} = SG$. The symbol $[\overset{?}{-}]$ indicates a $[-]$ which would undergo harmony provided no opaque vowel occurs downstream.

w	$+---\boxminus$	$+---$
$\overrightarrow{A}(w)$	$+\overset{?}{-}\overset{?}{-}\overset{?}{-}\boxminus$	$+\overset{?}{-}\overset{?}{-}\overset{?}{-}$
$\overleftarrow{B} \circ \overrightarrow{A}(w)$	$+---\boxminus$	$++++$

Table 5: Illustrations of the role of $[\overset{?}{-}]$ in the deterministic decomposition of SG=$\overleftarrow{B} \circ \overrightarrow{A}$.

7 Conclusion

The first suggestion that phonological processes have a tighter computational bound than "being regular" may come from (Mohri, 1997), buried on page 279. He writes without elaboration or citation "Most phonological and morphological rules correspond to p-subsequential relations." This study suggests that Mohri's assessment is largely correct, though the complete picture is more complicated than Mohri's offhand comment indicates.

The more complicated picture with respect to vowel harmony is expressed in Figure 6, which summarizes this paper's contributions. Traditional directional theories of vowel harmony express simpler generalizations than dominant/recessive/stem-control theories. It is our opinion that future work will likely show that even the weakly deterministic boundary surely separates the pathological patterns from the attested ones.

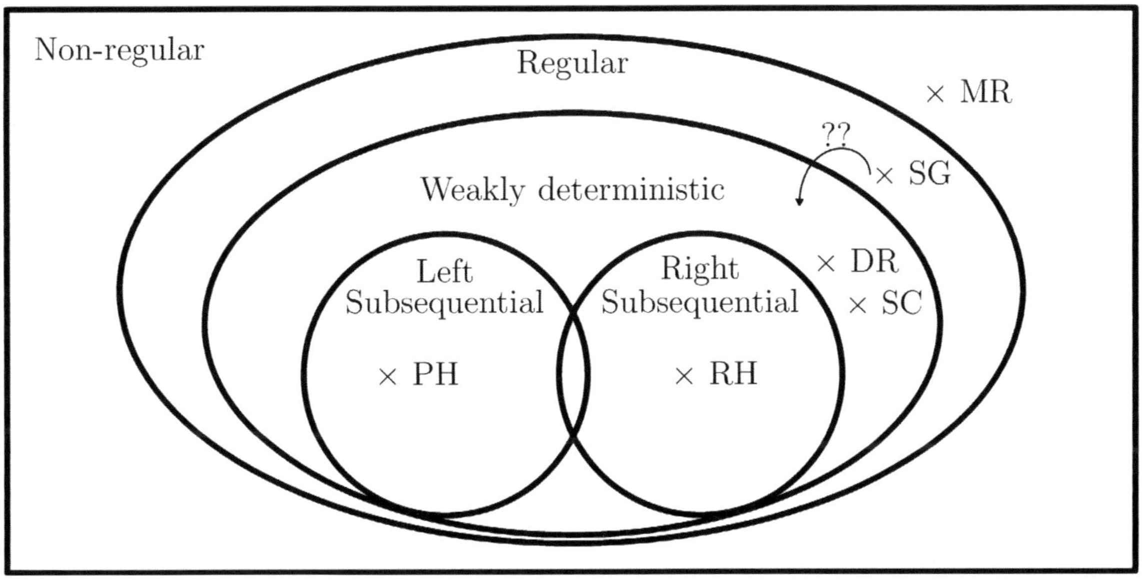

Figure 6: Hierarchies of transductions with the results of this paper shown. PH=progressive harmony, RH=regressive harmony, DR=dominant/recessive harmony, SC=stem control harmony, SG=sour grapes harmony, and MR=majority rules harmony.

Although the harmony patterns in this paper are all describable with same-length relations, we deliberately chose not to focus on the special properties same-length relations engender. This is largely because there are phonological processes such as epenthesis and deletion which are not same-length, and we would like our conclusions to hold for all phonological patterns. Nonetheless, future work which explores the same-lengthness aspect may lead to some interesting insights. One reviewer of this paper conjectured, for example, that if the same-length relations were coded as languages that they would then be k-reversible (Angluin, 1982).

With respect to learnability, total subsequential functions are identifiable in the limit from positive data (Oncina et al., 1993), though this algorithm appears to require data points unavailable in natural language corpora (Gildea and Jurafsky, 1996). Investigating subclasses of subsequential functions which cover attested phonological patterns may thus not only better characterize possible phonologies, but may also provide insights for learning (Chandlee and Koirala, 2013).

Finally, we believe Elgot and Mezei's theorem can shed new light on the old problem of abstractness in phonology (Hyman, 1970), and suspect a hierarchy of complexity depending on how much markup (either new symbols or with a length-increasing function) needs to be introduced in the intermediate alphabet to order to decompose a regular function into left and right subsequential ones. Computational work whose results should be more carefully investigated with this in mind include Kempe (2000) and Crespi Reghizzi and San Pietro (2012).

Acknowledgments

We thank the three anonymous reviewers for their insightful comments. We also thank the 2012-2013 computational linguistics group at the University of Delaware, especially Jim Rogers, for useful discussion.

References

Dana Angluin. 1982. Inference of reversible languages. *Journal for the Association of Computing Machinery*, 29(3):741–765.

Diana Archangeli and Douglas Pulleyblank. 1994. *Grounded Phonology*. Cambridge, MA: MIT Press.

Eric Baković. 2000. *Harmony, Dominance and Control*. Ph.D. thesis, Rutgers University.

Eric Baković. 2003. Vowel harmony and stem identity. In *San Diego Linguistic Papers*, pages 1–42. Department of Linguistics, UCSD, UC San Diego. `http://escholarship.org/uc/item/7zw206pt`.

Kenneth Beesley and Lauri Kartunnen. 2003. *Finite State Morphology*. CSLI Publications.

Michael Benedikt, Leonid Libkin, Thomas Schwentick, and Luc Segoufin. 2001. A model-theoretic approach to regular string relations. *Logic in Computer Science, Symposium on*, 0:0431.

Jean Berstel. 1979. *Transductions and Context-Free languages*. Teubner-Verlag.

Jane Chandlee and Jeffrey Heinz. 2012. Bounded copying is subsequential: Implications for metathesis and reduplication. In *Proceedings of the 12th Meeting of the ACL Special Interest Group on Computational Morphology and Phonology*, pages 42–51, Montreal, Canada, June. Association for Computational Linguistics.

Jane Chandlee and Cesar Koirala. 2013. Learning local phonological rules. Penn Linguistics Colloquium, March.

Jane Chandlee, Angeliki Athanasopoulou, and Jeffrey Heinz. 2012. Evidence for classifying metathesis patterns as subsequential. In *The Proceedings of the 29th West Coast Conference on Formal Linguistics*, pages 303–309. Cascillida Press.

Noam Chomsky and Morris Halle. 1968. *The Sound Pattern of English*. Harper & Row, New York.

Stefano Crespi Reghizzi and Pierluigi San Pietro. 2012. From regular to strictly locally testable languages. *International Journal of Foundations of Computer Science*, 23(08):1711–1727.

C. C. Elgot and J. E. Mezei. 1965. On relations defined by generalized finite automata. *IBM Journal of Research and Development*, 9(1):47–68.

Sara Finley. 2008. *The formal and cognitive restrictions on vowel harmony*. Ph.D. thesis, Johns Hopkins University, Baltimore, MD.

Brian Gainor, Regine Lai, and Jeffrey Heinz. 2012. Computational characterizations of vowel harmony patterns and pathologies. In *The Proceedings of the 29th West Coast Conference on Formal Linguistics*, pages 63–71.

Daniel Gildea and Daniel Jurafsky. 1996. Learning bias and phonological-rule induction. *Computational Linguistics*, 24(4).

Bruce Hayes. 2009. *Introductory Phonology*. Wiley-Blackwell.

Jeffrey Heinz and William Idsardi. 2011. Sentence and word complexity. *Science*, 333(6040):295–297, July.

Jeffrey Heinz and William Idsardi. 2013. What complexity differences reveal about domains in language. *Topics in Cognitive Science*, 5(1):111–131.

Jeffrey Heinz, Chetan Rawal, and Herbert G. Tanner. 2011. Tier-based strictly local constraints for phonology. In *Proceedings of the 49th Annual Meeting of the Association for Computational Linguistics*, pages 58–64, Portland, Oregon, USA, June. Association for Computational Linguistics.

Larry Hyman. 1970. How concrete is phonology? *Language*, 46(1):58–76.

Larry Hyman. 1975. *Phonology: Theory and Analysis*. Holt, Rinehart and Winston.

C. Douglas Johnson. 1972. *Formal Aspects of Phonological Description*. The Hague: Mouton.

Ronald Kaplan and Martin Kay. 1994. Regular models of phonological rule systems. *Computational Linguistics*, 20(3):331–378.

André Kempe. 2000. Reduction of intermediate alphabets in finite-state transducer cascades. *CoRR*, cs.CL/0010030.

Michael Kenstowicz. 1994. *Phonology in Generative Grammar*. Blackwell Publishers.

Kimmo Koskenniemi. 1983. Two-level morphology. Publication no. 11, Department of General Linguistics. Helsinki: University of Helsinki.

Martin Krämer. 2003. *Vowel Harmony and Correspondence Theory*. Berlin: Mouton de Gruyter.

Regine Lai. 2012. *Domain Specificity in Phonology*. Ph.D. thesis, University of Delaware.

Linda Lombardi. 1999. Positional faithfulness and voicing assimilation in Optimality Theory. *Natural Language and Linguistic Theory*, 17:267–302.

Mehryar Mohri. 1997. Finite-state transducers in language and speech processing. *Computational Linguistics*, 23(2):269–311.

José Oncina, Pedro García, and Enrique Vidal. 1993. Learning subsequential transducers for pattern recognition tasks. *IEEE Transactions on Pattern Analysis and Machine Intelligence*, 15:448–458, May.

Jaye Padgett. 1995. Partial class behavior and nasal place assimilation. In K. Suzuki and D. Elzinga, editors, *Proceedings of the 1995 Southwestern Workshop on Optimality Theory*.

Alan Prince and Paul Smolensky. 2004. *Optimality Theory: Constraint Interaction in Generative Grammar*. Blackwell Publishing.

Jason Riggle. 2004. *Generation, Recognition, and Learning in Finite State Optimality Theory*. Ph.D. thesis, University of California, Los Angeles.

Emmanuel Roche and Yves Schabes. *Finite-State Language Processing*. MIT Press.

James Rogers, Jeffrey Heinz, Margaret Fero, Jeremy Hurst, Dakotah Lambert, and Sean Wibel. to appear. Cognitive and sub-regular complexity. In *Proceedings of the 17th Conference on Formal Grammar*.

Jaques Sakarovitch. 2009. *Elements of Automata Theory*. Cambridge University Press. Translated by Reuben Thomas from the 2003 edition published by Vuibert, Paris.

Colin Wilson. 2003. Analyzing unbounded spreading with constraints: marks, targets, and derivations. Unpublished manuscript, UCLA.

Colin Wilson. 2004. Experimental investigation of phonological naturalness. In *Proceedings of WCCFL 22*, pages 534–546.

Anssi Yli-Jyrä and Kimmo Koskenniemi. 2006. Compiling generalized two-level rules and grammars. In *FinTAL*, volume 4139 of *Lecture Notes in Artificial Intelligence*, pages 174–185. Springer-Verlag, Berlin.

Learning Subregular Classes of Languages with Factored Deterministic Automata

Jeffrey Heinz
Dept. of Linguistics and Cognitive Science
University of Delaware
heinz@udel.edu

James Rogers
Dept. of Computer Science
Earlham College
jrogers@cs.earlham.edu

Abstract

This paper shows how factored finite-state representations of subregular language classes are identifiable in the limit from positive data by learners which are polytime iterative and optimal. These representations are motivated in two ways. First, the size of this representation for a given regular language can be exponentially smaller than the size of the minimal deterministic acceptor recognizing the language. Second, these representations (including the exponentially smaller ones) describe actual formal languages which successfully model natural language phenomenon, notably in the subfield of phonology.

1 Introduction

In this paper we show how to define certain subregular classes of languages which are identifiable in the limit from positive data (ILPD) by efficient, well-behaved learners with a lattice-structured hypothesis space (Heinz et al., 2012). It is shown that every finite *set* of DFAs defines such an ILPD class. In this case, each DFA can be viewed as one *factor* in the description of every language in the class. This factoring of language classes into multiple DFA can provide a *compact*, *canonical* representation of the grammars for *every* language in the class. Additionally, many subregular classes of languages can be learned by the above methods including the Locally k-Testable, Strictly k-Local, Piecewise k-Testable, and Strictly k-Piecewise languages (McNaughton and Papert, 1971; Rogers and Pullum, 2011; Rogers et al., 2010). From a linguistic (and cognitive) perspective, these subregular classes are interesting because they appear to be sufficient for modeling phonotactic patterns in human language (Heinz, 2010; Heinz et al., 2011; Rogers et al., to appear).

2 Preliminaries

For any function f and element a in the domain of f, we write $f(a)\downarrow$ if $f(a)$ is defined, $f(a)\downarrow = x$ if it is defined for a and its value is x, and $f(a)\uparrow$ otherwise. The range of f, the set of values f takes at elements for which it is defined, is denoted $\text{range}(f)$.

Σ^* and Σ^k denote all sequences of any finite length, and of length k, over a finite alphabet Σ. The empty string is denoted λ. A language L is a subset of Σ^*.

For all x, y belonging to a partially-ordered set $(S, \leq)$, if $x \leq z$ and $y \leq z$ then z is an *upper bound* of x and y. For all $x, y \in S$, the *least* upper bound (lub) $x \sqcup y = z$ iff $x \leq z$, $y \leq z$, and for all z' which upper bound x and y, it is the case that $z \leq z'$. An *upper semi-lattice* is a partially ordered set $(S, \leq)$ such that every subset of S has a lub. If S is finite, this is equivalent to the existence of $x \sqcup y$ for all $x, y \in S$.

A deterministic finite-state automaton (DFA) is a tuple $(Q, \Sigma, Q_0, F, \delta)$. The states of the DFA are Q; the input alphabet is Σ; the set of initial states is Q_0; the final states are F; and $\delta : Q \times \Sigma \to Q$ is the transition *function*.

We admit a set of initial states solely to accommodate the empty DFA, which has none. Deterministic automata never have more than one initial state. We will assume that, if the automaton is non-empty, then $Q_0 = \{q_0\}$;

The transition function's domain is extended to $Q \times \Sigma^*$ in the usual way.

The language of a DFA $\mathcal{A}$ is

$$L(\mathcal{A}) \stackrel{\text{def}}{=} \{w \in \Sigma^* \mid \delta(q_0, w)\downarrow \in F\}.$$

A DFA is *trim* iff it has no useless states:

$$(\forall q \in Q)[\ \exists w, v \in \Sigma^* \mid \delta(q_0, w)\downarrow = q \text{ and } \delta(q, v)\downarrow \in F].$$

Every DFA can be trimmed by eliminating useless states from Q and restricting the remaining components accordingly.

The empty DFA is $\mathcal{A}_\varnothing = (\varnothing, \Sigma, \varnothing, \varnothing, \varnothing)$. This is the minimal trim DFA such that $L(\mathcal{A}_\varnothing) = \varnothing$.

The DFA product of $\mathcal{A}_1 = (Q_1, \Sigma, Q_{01}, F_1, \delta_1)$ and $\mathcal{A}_2 = (Q_2, \Sigma, Q_{02}, F_2, \delta_2)$ is

$$\otimes(\mathcal{A}_1, \mathcal{A}_2) = (Q, \Sigma, Q_0, F, \delta)$$

where $Q = Q_1 \times Q_2$, $Q_0 = Q_{01} \times Q_{02}$, $F = F_1 \times F_2$ and

$$(\forall q \in Q)(\forall \sigma \in \Sigma)\big[\ \delta\big((q_1, q_2), \sigma\big) \stackrel{\text{def}}{=} (\delta_1(q_1, \sigma), \delta_2(q_2, \sigma))\ \big]$$

The DFA product of two DFA is also a DFA. It is not necessarily trim, but we will generally assume that in taking the product the result has been trimmed, as well.

The product operation is associative and commutative (up to isomorphism), and so it can be applied to a finite set S of DFA, in which case we write $\bigotimes S = \bigotimes_{\mathcal{A}\in S} \mathcal{A}$ (letting $\bigotimes\{\mathcal{A}\} = \mathcal{A}$). In this paper, grammars are finite *sequences* of DFAs $\vec{\mathcal{A}} = \langle \mathcal{A}_1 \cdots \mathcal{A}_n \rangle$ and we also use the $\bigotimes$ notation for the product of a finite sequence of DFAs: $\bigotimes \vec{\mathcal{A}} \stackrel{\text{def}}{=} \bigotimes_{\mathcal{A}\in\vec{\mathcal{A}}} \mathcal{A}$ and $L(\vec{\mathcal{A}}) \stackrel{\text{def}}{=} L(\bigotimes \vec{\mathcal{A}})$. Sequences are used instead of sets in order to match factors in two grammars. Let $\mathcal{DFA}$ denote the collection of finite sequences of DFAs.

Theorem 1 is well-known.

Theorem 1 *Consider a finite set S of DFA. Then* $L(\bigotimes_{\mathcal{A}\in S} \mathcal{A}) = \bigcap_{\mathcal{A}\in S} L(\mathcal{A})$.

An important consequence of Theorem 1 is that some languages are exponentially more compactly represented by their factors. The grammar $\vec{\mathcal{A}} = \langle \mathcal{A}_1 \cdots \mathcal{A}_n \rangle$ has $\sum_{1\le i\le n} \mathbf{card}(Q_i)$ states, whereas the trimmed $\bigotimes \vec{\mathcal{A}}$ can have as many as $\prod_{1\le i\le n} \mathbf{card}(Q_i) \in \Theta(\max_{1\le i\le n}(\mathbf{card}(Q_i))^n)$ states. An example of such a language is given in Section 4, Figures 1 and 2.

2.1 Identification in the limit

A *positive text* T for a language L is a total function $T : \mathbb{N} \to L \cup \{\#\}$ ($\#$ is a 'pause') such that range$(T) = L$ (i.e., for every $w \in L$ there is at least one $n \in \mathbb{N}$ for which $w = T(n)$). Let $T[i]$ denote the initial finite sequence $T(0), T(1) \dots T(i-1)$. Let SEQ denote the set of all finite initial portions of all positive texts for all possible languages. The *content* of an element $T[i]$ of SEQ is

$$content(T[i]) \stackrel{\text{def}}{=} \{w \in \Sigma^* \mid (\exists j \le i-1)[T(j) = w]\}.$$

In this paper, learning algorithms are programs: $\phi : SEQ \to \mathcal{DFA}$. A learner ϕ *identifies in the limit from positive texts* a collection of languages $\mathcal{L}$ if and only if for all $L \in \mathcal{L}$, for all positive texts T for L, there exists an $n \in \mathbb{N}$ such that

$$(\forall m \ge n)[\phi(T[m]) = \phi(T[n])] \text{ and } L(T[n]) = L$$

(see Gold (1967) and Jain et al. (1999)). A class of languages is ILPD iff it is identifiable in the limit by such a learner.

3 Classes of factorable-DFA languages

In this section, classes of factorable-DFA languages are introduced. The notion of sub-DFA is central to this concept. Pictorially, a sub-DFA is obtained from a DFA by removing zero or more states, transitions, and/or revoking the final status of zero or more final states.

Definition 1 *For any DFA $\mathcal{A} = (Q, \Sigma, Q_0, F, \delta)$, a DFA $\mathcal{A}' = (Q', \Sigma', Q'_0, F', \delta')$ is* sub-DFA *of $\mathcal{A}$, written $\mathcal{A}' \sqsubseteq \mathcal{A}$, if and only if $Q' \subseteq Q$, $\Sigma \subseteq \Sigma'$, $Q'_0 \subseteq Q_0$, $F' \subseteq F$, $\delta' \subseteq \delta$.*

The sub-DFA relation is extended to grammars (sequences of DFA). Let $\vec{\mathcal{A}} = \langle \mathcal{A}_1 \cdots \mathcal{A}_n \rangle$ and $\vec{\mathcal{A}}' = \langle \mathcal{A}'_1 \cdots \mathcal{A}'_n \rangle$.

Then $\vec{\mathcal{A}}' \sqsubseteq \vec{\mathcal{A}} \Leftrightarrow (\forall 0 \le i \le n)[\mathcal{A}'_i \sqsubseteq \mathcal{A}_i]$.

Clearly, if $\mathcal{A}' \sqsubseteq \mathcal{A}$ then $L(\mathcal{A}') \subseteq L(\mathcal{A})$.

Every grammar $\vec{\mathcal{A}}$ determines a class of languages: those recognized by a sub-grammar of $\vec{\mathcal{A}}$. Our interest is not in $L(\vec{\mathcal{A}})$, itself. Indeed, this will generally be Σ^*. Rather, our interest is in identifying languages relative to the class of languages recognizable by sub-grammars of $\vec{\mathcal{A}}$.

Definition 2 *Let $\mathcal{G}(\vec{\mathcal{A}}) \stackrel{def}{=} \{\vec{\mathcal{B}} \mid \vec{\mathcal{B}} \sqsubseteq \vec{\mathcal{A}}\}$, the class of grammars that are sub-grammars of $\vec{\mathcal{A}}$.*

Let $\mathcal{L}(\vec{\mathcal{A}}) \stackrel{def}{=} \{L(\vec{\mathcal{B}}) \mid \vec{\mathcal{B}} \sqsubseteq \vec{\mathcal{A}}\}$, the class of languages recognized by sub-grammars of $\vec{\mathcal{A}}$.

A class of languages is a factorable-DFA class iff it is $\mathcal{L}(\vec{\mathcal{A}})$ for some $\vec{\mathcal{A}}$.

The set $\mathcal{G}(\vec{\mathcal{A}})$ is necessarily finite, since $\vec{\mathcal{A}}$ is, so every class $\mathcal{L}(\vec{\mathcal{A}})$ is trivially ILPD by a learning algorithm that systematically rules out grammars that are incompatible with the text, but this naïve algorithm is prohibitively inefficient. Our goal is to establish that the efficient general learning algorithm given by Heinz et al. (2012) can be applied to every class of factorable-DFA languages, and that this class includes many of the well-known sub-regular language classes as well as classes that are, in a particular sense, mixtures of these.

4 A motivating example

This section describes the Strictly 2-Piecewise languages, which motivate the factorization that is at the heart of this analysis. Strictly Piecewise (SP) languages are characterized in Rogers et al. (2010) and are a special subclass of the Piecewise Testable languages (Simon, 1975).

Every SP language is the intersection of a finite set of complements of principal shuffle ideals:

$$L \in \text{SP} \overset{\text{def}}{\Longleftrightarrow} L = \bigcap_{w \in S} [\overline{\text{SI}(w)}], \quad S \text{ finite}$$

where

$$\begin{aligned}\text{SI}(w) \overset{\text{def}}{=} \{v \in \Sigma^* \mid w = \sigma_1 \cdots \sigma_k \text{ and} \\ (\exists v_0, \ldots, v_k \in \Sigma^*)[v = v_0 \cdot \sigma_1 \cdot v_1 \cdots \sigma_k \cdot v_k]\}\end{aligned}$$

So $v \in \text{SI}(w)$ iff w occurs as a subsequence of v and $L \in \text{SP}$ iff there is a finite set of strings for which L includes all and only those strings that do not include those strings as subsequences. We say that L is *generated* by S. It turns out that SP is exactly the class of languages that are closed under subsequence.

A language is SP_k iff it is generated by a set of strings each of which is of length less than or equal to k. Clearly, every SP language is SP_k for some k and $\text{SP} = \bigcup_{1 \leq k \in \mathbb{N}} [\text{SP}_k]$.

If $w \in \Sigma^*$ and $|w| = k$, then $\overline{\text{SI}(w)} = L(\mathcal{A}_{\overline{w}})$ for a DFA $\mathcal{A}_{\overline{w}}$ with no more than k states. For example, if $k = 2$ and $\Sigma = \{a, b, c\}$ and, hence, $w \in \{a, b, c\}^2$, then the minimal trim DFA recognizing $\overline{\text{SI}(w)}$ will be a sub-DFA (in which one of the transitions from the σ_1 state has been removed) of one of the three DFA of Figure 1.

Figure 1 shows $\vec{\mathcal{A}} = \langle A_a, A_b, A_c \rangle$, where $\Sigma = \{a, b, c\}$ and each A_σ is a DFA accepting Σ^* whose states distinguish whether σ has yet occurred. Figure 2 shows $\bigotimes \vec{\mathcal{A}}$.

Note that every SP_2 language over $\{a, b, c\}$ is $L(\vec{\mathcal{B}})$ for some $\vec{\mathcal{B}} \sqsubseteq \vec{\mathcal{A}}$. The class of grammars of $\mathcal{G}(\vec{\mathcal{A}})$ recognize a slight extension of SP_2 over $\{a, b, c\}$ (which includes 1-Reverse Definite languages as well).

Observe that 6 states are required to describe $\vec{\mathcal{A}}$ but 8 states are required to describe $\bigotimes \vec{\mathcal{A}}$. Let $\vec{\mathcal{A}}_\Sigma$ be the sequence of DFA with one DFA for each letter in Σ, as in Figure 1. As $\textbf{card}(\Sigma)$ increases the number of states of $\vec{\mathcal{A}}_\Sigma$ is $2 \times \textbf{card}(\Sigma)$ but the number of states in $\bigotimes \vec{\mathcal{A}}_\Sigma$ is $2^{\textbf{card}(\Sigma)}$. The number of states in the product, in this case, is exponential in the number of its factors.

The Strictly 2-Piecewise languages are currently the strongest computational characterization[1] of long-distance phonotactic patterns in human languages (Heinz, 2010). The size of the phonemic inventories[2] in the world's languages ranges from 11 to 140 (Maddieson, 1984). English has about 40, depending on the dialect. With an alphabet of that size $\vec{\mathcal{A}}_\Sigma$ would have 80 states, while $\bigotimes \vec{\mathcal{A}}_\Sigma$ would have $2^{40} \approx 1 \times 10^{12}$ states. The fact that there are about 10^{11} neurons in human brains (Williams and Herrup, 1988) helps motivate interest in the more compact, parallel representation given by $\vec{\mathcal{A}}_\Sigma$ as opposed to the singular representation of the DFA $\bigotimes \vec{\mathcal{A}}_\Sigma$.

5 Learning factorable classes of languages

In this section, classes of factorable-DFA languages are shown to be analyzable as finite lattice spaces. By Theorem 6 of Heinz et al. (2012), every such class of languages can be identified in the limit from positive texts.

Definition 3 (Joins) *Let*

$$\begin{aligned} \mathcal{A} &= (Q, \Sigma, Q_0, F, \delta), \\ \mathcal{A}_1 &= (Q_1, \Sigma, Q_{01}, F_1, \delta_1) \sqsubseteq \mathcal{A} \\ \text{and} & \\ \mathcal{A}_2 &= (Q_2, \Sigma, Q_{02}, F_2, \delta_2) \sqsubseteq \mathcal{A}. \end{aligned}$$

The join *of* $\mathcal{A}_1$ *and* $\mathcal{A}_2$ *is*

$$\mathcal{A}_1 \sqcup \mathcal{A}_2 \overset{\text{def}}{=} (Q_1 \cup Q_2, \Sigma, Q_{01} \cup Q_{02}, F_1 \cup F_2, \delta_1 \cup \delta_2).$$

Similarly, for all $\vec{\mathcal{A}} = \langle \mathcal{A}_1 \cdots \mathcal{A}_n \rangle$ *and* $\vec{\mathcal{B}} = \langle \mathcal{B}_1 \cdots \mathcal{B}_n \rangle \sqsubseteq \vec{\mathcal{A}}$, $\vec{\mathcal{C}_2} = \langle \mathcal{C}_1 \cdots \mathcal{C}_n \rangle \sqsubseteq \vec{\mathcal{A}}$, *the* join *of and* $\vec{\mathcal{B}}$ *and* $\vec{\mathcal{C}}$ *is* $\vec{\mathcal{B}} \sqcup \vec{\mathcal{C}} \overset{\text{def}}{=} \langle \mathcal{B}_1 \sqcup \mathcal{C}_1 \cdots \mathcal{B}_n \sqcup \mathcal{C}_n \rangle$.

Note that the join of two sub-DFA of $\mathcal{A}$ is also a sub-DFA of $\mathcal{A}$. Since $\mathcal{G}(\vec{\mathcal{A}})$ is finite, binary join suffices to define join of any set of sub-DFA of a given DFA (as iterated binary joins). Let $\bigsqcup[S]$ be the join of S, a set of sub-DFAs of some $\mathcal{A}$ (or $\vec{\mathcal{A}}$).

[1] See Heinz et al. (2011) for competing characterizations.

[2] The mental representations of speech sounds are called phonemes, and the phonemic inventory is the set of these representations (Hayes, 2009).

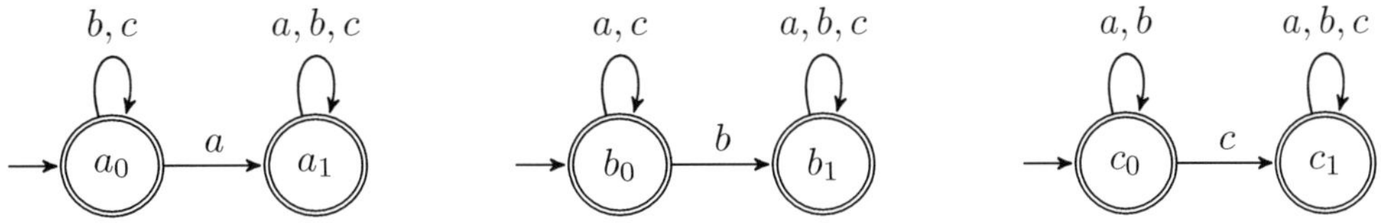

Figure 1: The sequence of DFA $\vec{\mathcal{A}} = \langle A_a, A_b, A_c \rangle$, where $\Sigma = \{a, b, c\}$ and each A_σ accepts Σ^* and whose states distinguish whether σ has yet occurred.

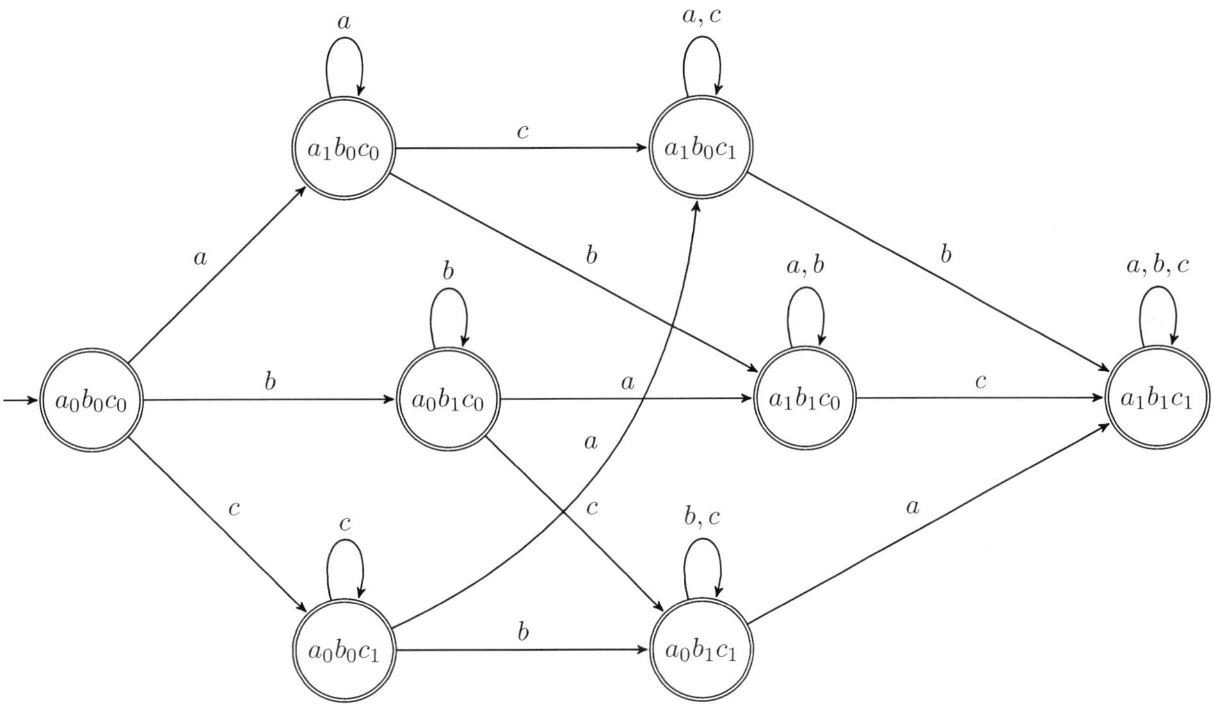

Figure 2: The product $\bigotimes \langle A_a, A_b, A_c \rangle$.

Lemma 1 *The set of sub-DFA of a DFA $\mathcal{A}$, ordered by $\sqsubseteq$, $(\{\mathcal{B} \mid \mathcal{B} \sqsubseteq \mathcal{A}\}, \sqsubseteq)$, is an upper semi-lattice with the least upper bound of a set of S sub-DFA of $\mathcal{A}$ being their join.*

Similarly the set of sub-grammars of a grammar $\vec{\mathcal{A}}$, ordered again by $\sqsubseteq$, $(\{\vec{\mathcal{B}} \sqsubseteq \vec{\mathcal{A}}\}, \sqsubseteq)$, is an upper semi-lattice with the least upper bound of a set of sub-grammars of $\vec{\mathcal{A}}$ being their join.[3]

This follows from the fact that $Q_1 \cup Q_2$ (similarly $F_1 \cup F_2$ and $\delta_1 \cup \delta_2$) is the lub of Q_1 and Q_2 (etc.) in the lattice of sets ordered by subset.

5.1 Paths and Chisels

Definition 4 *Let $\mathcal{A} = (Q, \Sigma, \{q_0\}, F, \delta)$ be a non-empty DFA and $w = \sigma_0\sigma_1 \cdots \sigma_n \in \Sigma^*$.*

If $\delta(q_0, w)\downarrow$, the path *of w in $\mathcal{A}$ is the sequence*

$$\pi(\mathcal{A}, w) \stackrel{def}{=} \langle (q_0, \sigma_0), \ldots, (q_n, \sigma_n), (q_{n+1}, \lambda) \rangle$$

where $(\forall 0 \leq i \leq n)[q_{i+1} = \delta(q_i, \sigma_i)]$.

If $\delta(q_0, w)\uparrow$ then $\pi(\mathcal{A}, w)\uparrow$.

If $\pi(\mathcal{A}, w)\downarrow$, let $Q_{\pi(\mathcal{A},w)}$ denote set of states it traverses, $\delta_{\pi(\mathcal{A},w)}$ denote the the transitions it traverses, and let $F_{\pi(\mathcal{A},w)} = \{q_{n+1}\}$.

Next, for any DFA $\mathcal{A}$, and any $w \in L(\mathcal{A})$, we define the *chisel* of w given $\mathcal{A}$ to be the sub-DFA of $\mathcal{A}$ that exactly encompasses the path etched out in $\mathcal{A}$ by w.

Definition 5 *For any non-empty DFA $\mathcal{A} = (Q, \Sigma, \{q_0\}, F, \delta)$ and all $w \in \Sigma^*$, if $w \in L(\mathcal{A})$, then the* chisel *of w given $\mathcal{A}$ is the sub-DFA*

$$C_{\mathcal{A}}(w) = (Q_{\pi(\mathcal{A},w)}, \Sigma, \{q_0\}, F_{\pi(\mathcal{A},w)}, \delta_{\pi(\mathcal{A},w)}).$$

If $w \notin L(\mathcal{A})$, then $C_{\mathcal{A}}(w) = \mathcal{A}_{\varnothing}$.

Consider any $\vec{\mathcal{A}} = \langle \mathcal{A}_1 \cdots \mathcal{A}_n \rangle$ and any word $w \in \Sigma^*$. The chisel *of w given $\vec{\mathcal{A}}$ is $C_{\vec{\mathcal{A}}}(w) = \langle C_{\mathcal{A}_1}(w) \cdots C_{\mathcal{A}_n}(w) \rangle$.*

Observe that $C_{\mathcal{A}}(w) \sqsubseteq \mathcal{A}$ for all words w and all $\mathcal{A}$, and that $C_{\mathcal{A}}(w)$ is trim.

Using the join, the domain of the chisel is extended to sets of words: $C_{\vec{\mathcal{A}}}(S) = \bigsqcup_{w \in S} C_{\vec{\mathcal{A}}}(w)$. Note that $\{C_{\vec{\mathcal{A}}}(w) \mid w \in \Sigma^*\}$ is finite, since $\{\vec{\mathcal{B}} \mid \vec{\mathcal{B}} \sqsubseteq \vec{\mathcal{A}}\}$ is.

Theorem 2 *For any grammar $\vec{\mathcal{A}}$, let $\mathcal{C}(\vec{\mathcal{A}}) = \{C_{\vec{\mathcal{A}}}(S) \mid S \subseteq \Sigma^*\}$. Then $(\mathcal{C}(\vec{\mathcal{A}}), \sqsubseteq)$ is an upper semi-lattice with the lub of two elements given by the join $\sqcup$.*

[3]These are actually complete finite lattices, but we are interested primarily in the joins.

Proof This follows immediately from the finiteness of $\{C_{\vec{\mathcal{A}}}(w) \mid w \in \Sigma^*\}$ and Lemma 1. □

Lemma 2 *For all $\mathcal{A} = (Q, \Sigma, Q_0, F, \delta)$, there is a finite set $S \subset \Sigma^*$ such that $\bigsqcup_{w \in S} C_{\mathcal{A}}(w) = \mathcal{A}$. Similarly, for all $\vec{\mathcal{A}} = \langle \mathcal{A}_1 \cdots \mathcal{A}_n \rangle$, there is a finite set $S \subset \Sigma^*$ such that $C_{\vec{\mathcal{A}}}(S) = \vec{\mathcal{A}}$.*

Proof If $\mathcal{A}$ is empty, then clearly $S = \varnothing$ suffices. Henceforth consider only nonempty $\mathcal{A}$.

For the first statement, let S be the set of $u\sigma v$ where, for each $q \in Q$ and for each $\sigma \in \Sigma$, $\delta(q_0, u)\downarrow = q$ and $\delta(\delta(q, \sigma), v)\downarrow \in F$ such that $u\sigma v$ has minimal length. By construction, S is finite. Furthermore, for every state and every transition in $\mathcal{A}$, there is a word in S whose path touches that state and transition. By definition of $\sqcup$ it follows that $C_{\mathcal{A}}(S) = \mathcal{A}$.

For proof of the second statement, for each $\mathcal{A}_i$ in $\vec{\mathcal{A}}$, construct S_i as stated and take their union. □

Heinz et al. (2012) define lattice spaces. For an upper semi-lattice V and a function $f : \Sigma^* \to V$ such that f and $\sqcup$ are (total) computable, (V, f) is called a *Lattice Space (LS)* iff, for each $v \in V$, there exists a finite $D \subseteq range(f)$ with $\bigsqcup D = v$.

Theorem 3 *For all grammars $\vec{\mathcal{A}} = \langle \mathcal{A}_1 \cdots \mathcal{A}_n \rangle$, $(\mathcal{C}(\vec{\mathcal{A}}), C_{\vec{\mathcal{A}}})$ is a lattice space.*

Proof For all $\vec{\mathcal{A}}' \in \mathcal{C}(\vec{\mathcal{A}})$, by Lemma 2, there is a finite $S \subseteq \Sigma^*$ such that $\bigsqcup_{w \in S} C_{\vec{\mathcal{A}}}(w) = \vec{\mathcal{A}}'$. □

For Heinz et al. (2012), elements of the lattice are grammars. Likewise, here, each grammar $\vec{\mathcal{A}} = \langle \mathcal{A}_1 \cdots \mathcal{A}_n \rangle$ defines a lattice whose elements are its sub-grammars. Heinz et al. (2012) associate the *languages* of a grammar v in a lattice space (V, f) with $\{w \in \Sigma^* \mid f(w) \sqsubseteq v\}$. This definition coincides with ours: for any element $\vec{\mathcal{A}}'$ of $\mathcal{C}(\vec{\mathcal{A}})$ (note $\vec{\mathcal{A}}' \sqsubseteq \vec{\mathcal{A}}$), a word w belongs to $L(\vec{\mathcal{A}}')$ if and only if $C_{\vec{\mathcal{A}}}(w)$ is a sub-DFA of $\vec{\mathcal{A}}'$. The *class of languages* of a LS is the collection of languages obtained by every element in the lattice. For every LS $(\mathcal{C}(\vec{\mathcal{A}}), C_{\vec{\mathcal{A}}})$, we now define a learner ϕ according to the construction in Heinz et al. (2012): $\forall T \in SEQ$, $\phi(T) = \bigsqcup_{w \in content(T)} C_{\vec{\mathcal{A}}}(w)$.

Let $\mathcal{L}_{(\mathcal{C}(\vec{\mathcal{A}}), C_{\vec{\mathcal{A}}})}$ denote the class of languages associated with the LS in Theorem 3. According to Heinz et al. (2012, Theorem 6), the learner ϕ identifies $\mathcal{L}_{(\mathcal{C}(\vec{\mathcal{A}}), C_v A)}$ in the limit from positive data. Furthermore, ϕ is *polytime iterative*, i.e can compute the next hypothesis in polytime from the previous hypothesis alone, and optimal in the sense that no other learner converges more quickly on languages in $\mathcal{L}_{(\mathcal{C}(\vec{\mathcal{A}}),C_G)}$. In addition, this learner is *globally-consistent* (every hypothesis covers the data seen so far), *locally-conservative* (the hypothesis never changes unless the current datum is not consistent with the current hypothesis), *strongly-monotone* (the current hypothesis is a superset of all prior hypotheses), and *prudent* (it never hypothesizes a language that is not in the target class). Formal definitions of these terms are given in Heinz et al. (2012) and can also be found elsewhere, e.g. Jain et al. (1999).

6 Complexity considerations

The space of sub-grammars of a given sequence of DFAs is necessarily finite and, thus, identifiable in the limit from positive data by a naïve learner that simply enumerates the space of grammars. The lattice learning algorithm has better efficiency because it works bottom-up, extending the grammar minimally, at each step, with the chisel of the current string of the text. The lattice learner never explores any part of the space of grammars that is not a sub-grammar of the correct one and, as it never moves down in the lattice, it will skip much of the space of grammars that are sub-grammars of the correct one. The space it explores will be minimal, given the text it is running on. Generalization is a result of the fact that in extending the grammar for a string the learner adds its entire Nerode equivalence class to the language.

The time complexity of either learning or recognition with the factored automata may actually be somewhat worse than the complexity of doing so with its product. Computing the chisel of a string w in the product machine of Figure 2 is $\Theta(|w|)$, while in the factored machine of Figure 1 one must compute the chisel in each factor and its complexity is, thus, $\Theta(|w|\,\mathbf{card}(\Sigma)^{k-1})$. But Σ and k are fixed for a given factorization, so this works out to be a constant factor.

Where the factorization makes a substantial difference is in the number of features that must be learned. In the factored grammar of the example, the total number of states plus edges is $\Theta(k\mathbf{card}(\Sigma)^{k-1})$, while in its product it is $\Theta(2^{\mathbf{card}(\Sigma)^{k-1}})$. This represents an exponential improvement in the space complexity of the factored grammar.

Every DFA can be factored in many ways, but the factorizations do not necessarily provide an asymptotically significant improvement in space complexity. The canonical contrast is between sequences of automata $\langle \mathcal{A}_1, \ldots, \mathcal{A}_n \rangle$ that count modulo some sequence of $m_i \in \mathbb{N}$. If the m_i are pairwise prime, the product will require $\prod_{1 \leq i \leq n}[m_i] = \Theta((\max_i[m_i])^n)$ states. If on the other hand, they are all multiples of each other it will require just $\Theta(\max_i[m_i])$.

7 Examples

The fact that the class of SP_2 languages is efficiently identifiable in the limit from positive data is neither surprising or new. The obvious approach to learning these languages simply accumulates the set of pairs of symbols that occur as subsequences of the strings in the text and builds a machine that accepts all and only those strings in which no other such pairs occur. This, in fact, is essentially what the lattice learner is doing.

What is significant is that the lattice learner provides a general approach to learning any language class that can be captured by a factored grammar and, more importantly, any class of languages that are intersections of languages that are in classes that can be captured this way.

Factored grammars in which each factor recognizes Σ^*, as in the case of Figure 1, are of particular interest. Every sub-Star-Free class of languages in which the parameters of the class (k, for example) are fixed can be factored in this way.[4] If the parameters are not fixed and the class of languages is not finite, none of these classes can be identified in the limit from positive data at all.[5] So this approach is potentially useful at least for all sub-Star-Free classes. The learners for non-strict classes are practical, however, only for small values of the parameters. So that leaves the Strictly Local SL_k and Strictly Piecewise SP_k languages as the obvious targets.

The SL_k languages are those that are determined by the substrings of length no greater than k that occur within the string (including endmark-

[4]We conjecture that there is a parameterized class of languages that is equivalent to the Star-Free languages, which would make that class learnable in this way as well.

[5]For most of these classes, including the Definite, Reverse-Definite and Strictly Local classes and their super classes, this is immediate from the fact that they are super-finite. SP, on the other hand, is not super-finite (since it does not include all finite languages) but nevertheless, it is not IPLD.

ers). These can be factored on the basis of those substrings, just as the SP_k languages can, although the construction is somewhat more complex. (See the Knuth-Morris-Pratt algorithm (Knuth et al., 1977) for a way of doing this.) But SL_k is a case in which there is no complexity advantage in factoring the DFA. This is because every SL_k language is recognized by a DFA that is a Myhill graph: with a state for each string of $\Sigma^{<k}$ (i.e., of length less than k). Such a graph has $\Theta(\mathbf{card}(\Sigma)^{k-1})$ states, asymptotically the same as the number of states in the factored grammar, which is actually marginally worse.

Therefore, factored SL_k grammars are not, in themselves, interesting. But they are interesting as factors of other grammars. Let $(SL + SP)_{k,l}$ (resp. $(LT + SP)_{k,l}$, $(SL + PT)_{k,l}$) be the class of languages that are intersections of SL_k and SP_l (resp. LT_k and SP_l, SL_k and PT_l) languages. Where LT (PT) languages are determined by the *set* of substrings (subsequences) that occur in the string (see Rogers and Pullum (2011) and Rogers et al. (2010)).

These classes capture co-occurrence of local constraints (based on adjacency) and long-distance constraints (based on precedence). These are of particular interest in phonotactics, as they are linguistically well-motivated approaches to modeling phonotactics and they are sufficiently powerful to model most phonotactic patterns. The results of Heinz (2007) and Heinz (2010) strongly suggest that nearly all segmental patterns are $(SL + SP)_{k,l}$ for small k and l. Moreover, roughly 72% of the stress patterns that are included in Heinz's database (Heinz, 2009; Phonology Lab, 2012) of patterns that have been attested in natural language can be modeled with SL_k grammars with $k \leq 6$. Of the rest, all but four are $LT_1 + SP_4$ and all but two are $LT_2 + SP_4$. Both of these last two are properly regular (Wibel et al., in prep).

8 Conclusion

We have shown how subregular classes of languages can be learned over factored representations, which can be exponentially more compact than representations with a single DFA. Essentially, words in the data presentation are passed through each factor, "activating" the parts touched. This approach immediately allows one to naturally "mix" well-characterized learnable subregular classes in such a way that the resulting language class is also learnable. While this mixing is partly motivated by the different kinds of phonotactic patterns in natural language, it also suggests a very interesting theoretical possibility. Specifically, we anticipate that the right parameterization of these well-studied subregular classes will cover the class of star-free languages. Future work could also include extending the current analysis to factoring stochastic languages, perhaps in a way that connects with earlier research on factored HMMs (Ghahramani and Jordan, 1997).

Acknowledgments

This paper has benefited from the insightful comments of three anonymous reviewers, for which the authors are grateful. The authors also thank Jie Fu and Herbert G. Tanner for useful discussion. This research was supported by NSF grant 1035577 to the first author, and the work was completed while the second author was on sabbatical at the Department of Linguistics and Cognitive Science at the University of Delaware.

References

Zoubin Ghahramani and Michael I. Jordan. 1997. Factorial hidden markov models. *Machine Learning*, 29(2):245–273.

E.M. Gold. 1967. Language identification in the limit. *Information and Control*, 10:447–474.

Bruce Hayes. 2009. *Introductory Phonology*. Wiley-Blackwell.

Jeffrey Heinz, Chetan Rawal, and Herbert G. Tanner. 2011. Tier-based strictly local constraints for phonology. In *Proceedings of the 49th Annual Meeting of the Association for Computational Linguistics*, pages 58–64, Portland, Oregon, USA, June. Association for Computational Linguistics.

Jeffrey Heinz, Anna Kasprzik, and Timo Kötzing. 2012. Learning with lattice-structured hypothesis spaces. *Theoretical Computer Science*, 457:111–127, October.

Jeffrey Heinz. 2007. *The Inductive Learning of Phonotactic Patterns*. Ph.D. thesis, University of California, Los Angeles.

Jeffrey Heinz. 2009. On the role of locality in learning stress patterns. *Phonology*, 26(2):303–351.

Jeffrey Heinz. 2010. Learning long-distance phonotactics. *Linguistic Inquiry*, 41(4):623–661.

Sanjay Jain, Daniel Osherson, James S. Royer, and Arun Sharma. 1999. *Systems That Learn: An Introduction to Learning Theory (Learning, Development and Conceptual Change)*. The MIT Press, 2nd edition.

Donald Knuth, James H Morris, and Vaughn Pratt. 1977. Fast pattern matching in strings. *SIAM Journal on Computing*, 6(2):323–350.

Ian Maddieson. 1984. *Patterns of Sounds*. Cambridge University Press, Cambridge, UK.

Robert McNaughton and Seymour Papert. 1971. *Counter-Free Automata*. MIT Press.

UD Phonology Lab. 2012. UD phonology lab stress pattern database. `http://phonology.cogsci.udel.edu/dbs/stress`. Accessed December 2012.

James Rogers and Geoffrey Pullum. 2011. Aural pattern recognition experiments and the subregular hierarchy. *Journal of Logic, Language and Information*, 20:329–342.

James Rogers, Jeffrey Heinz, Gil Bailey, Matt Edlefsen, Molly Visscher, David Wellcome, and Sean Wibel. 2010. On languages piecewise testable in the strict sense. In Christian Ebert, Gerhard Jäger, and Jens Michaelis, editors, *The Mathematics of Language*, volume 6149 of *Lecture Notes in Artifical Intelligence*, pages 255–265. Springer.

James Rogers, Jeffrey Heinz, Margaret Fero, Jeremy Hurst, Dakotah Lambert, and Sean Wibel. to appear. Cognitive and sub-regular complexity. In *Proceedings of the 17th Conference on Formal Grammar*.

Imre Simon. 1975. Piecewise testable events. In *Automata Theory and Formal Languages: 2nd Grammatical Inference conference*, pages 214–222, Berlin. Springer-Verlag.

Sean Wibel, James Rogers, and Jeffery Heinz. Factoring of stress patterns. In preparation.

R.W. Williams and K. Herrup. 1988. The control of neuron number. *Annual Review of Neuroscience*, 11:423–453.

Structure Learning in Weighted Languages

András Kornai, Attila Zséder, Gábor Recski
HAS Computer and Automation Research Institute
H-1111 Kende u 13-17, Budapest
{kornai,zseder,recski}@sztaki.mta.hu

Abstract

We present Minimum Description Length techniques for learning the structure of weighted languages. MDL is already widely used both for segmentation and classification tasks, and here we show it can be used to formalize further important tools in the descriptive linguists' toolbox, including the distinction between accidental and systematic gaps in the data, the detection of ambiguity, the selective discarding of data, and the merging of categories.

Introduction

The Minimum Description Length (MDL, see Rissanen 1978) framework is primarily about data compression: if we are given some data D, our goal is to find a model $\mathcal{M}$, and a correction term $\mathcal{E}$, such that the model output and the correction term together describe the data, and transmitting $\mathcal{M}$ and $\mathcal{E}$ takes fewer bits than transmitting any competing $\mathcal{M}'$ and $\mathcal{E}'$.

From the very beginning, starting with Pāṇini, linguists have put a premium on brevity. The hope is that the shortest theory is the best theory (see Vitanyi and Li 2000), at least if we are willing to posit a theory of Universal Grammar (UG) that will let us specify $\mathcal{M}$ briefly, since we can assume UG to be amortized over many languages.

In this paper we study the problem of compressing weighted languages by presenting them via weighted finite state automata (WFSA). The theoretical approach we discuss here has a long history: the founding paper of Kolmogorov complexity, Solomonoff (1964), already studied the problem of inferring a grammar from data, and Grünwald (1996) uses MDL to infer CFGs from corpora, there conceived of as long strings over a finite alphabet. It is fair to say that this theory has not had much impact on computational practice, where grammatical inference is dominated by the standard n-gram based language modeling methods, see Jelinek (1997) for an excellent summary of the basic ideas and techniques, most of which are still in wide use.

While the two approaches may coincide in certain cases (see Grünwald 1996), and in theory n-gram models are just a special case of the general WFSA, in practice they are divided by a fundamental difference in modeling unseen data. From an engineering standpoint, Church et al (2007) are entirely right in saying:

> No matter how much data we have, we never have enough. Nothing has zero probability.

Linguists, starting perhaps with Chomsky (1965), draw a bright line between *accidentally* and *systematically* missing data, and would prefer to restrict backoff techniques to the accidental gaps. The distinction is often lost in applied work, because the models need to be built in a noisy environment, where frequent typos like **teh* and similar performance errors can easily overwhelm genuine items like *boisterous* or *mopeds* by an order of magnitude or even more. In the eyes of many linguists, this observation alone is sufficient to rob probabilistic models of grammatical content, since this makes it impossible to define a single threshold g such that all and only strings with weight greater than g are grammatical.

Aside from this subtle but important distinction between accidental and systematic gaps, both kinds of language modeling can be cast in the same formal terms: we fit a model $\mathcal{M}$ that minimizes some function E (typically, the squared sum) of the error $\mathcal{E}$. Obviously, the more parameters $\mathcal{M}$ has, the better fit we can obtain. Much of contemporary computational linguistics follows the route of training simple models such as Hidden Markov Models (HMMs) and probabilistic context-free grammars (PCFGs) with very many parameters, and stops adding more only when compressing the memory footprint is of paramount importance. As (Church et al., 2007) notes, applications like the contextual speller of Microsoft Office simply could not ship without keeping the language model within reasonable size limits. In such cases, we are quite willing to trade in E for gains in the size M of $\mathcal{M}$, and considerations of optimizing the sum of the two are simply irrelevant.

In contrast, our strategy is to search for model which measures both $\mathcal{M}$ and $\mathcal{E}$ in bits, and optimizes the sum $M + E$, not because we put such a premium on data compression, but rather because we follow in Pāṇini's footsteps. Our goal is finding *structural* models capable of distinguishing structurally excluded (ungrammatical) strings like *furiously sleep ideas green colorless* from low probability but grammatical strings like *colorless green ideas sleep furiously* (Pereira, 2000). For this more ambitious goal comparing models with different number of parameters is a key issue, and this is precisely where MDL is helpful.

The rest of this Introduction provides the basic definitions, notation, and terminology, all fairly standard except for the use of Moore rather than Mealy machines – the significance of this choice will be discussed in Section 2. In Section 1 we bring a fundamental idea of signal processing, *quantization error*, to bear on the problem of model selection, illustrating the issue on a real example, the proquant system of Hungarian. In Section 2 we show how one of the most powerful tools at disposal of the linguist, *ambiguity*, can be detected by MDL, bringing another standard idea, *signal to noise ratio* to bear. In Section 3 we discuss another real example, Hungarian morphotactics, and show that two methods widely (but shamefacedly) used in practice, discarding data and merging descriptive categories, can be used on a principled basis within MDL. Our goal is to show that by consistent application of MDL principles we can automatically set up the kind of models that linguists would set up. Ultimately, both man and machine work toward the same goal, optimization of grammar elegance or, what is the same, brevity.

Definition 1. Given some finite alphabet Σ, a *weighted language* p over this alphabet is defined as a mapping $p : \Sigma^* \rightarrow \mathbb{R}$ taking non-negative values such that $\sum_{\alpha \in \Sigma^*} p(\alpha) = 1$. This is less general than the standard notion of noncommutative power series with weights taken in arbitrary semirings (Eilenberg 1974, Salomaa 1978) but will suffice here. The stringset $\{\alpha | p(\alpha) > 0\}$ is called the *support* of p and will be denoted by $S(p)$.

Definition 2. Given two weighted languages p and q, we say the Kullback-Leibler (KL) *approximation error* Q of q relative to p is $\sum_{\alpha \in S(q)} p(\alpha) \log(p(\alpha)/q(\alpha))$. The *entropy* of p is defined as $-\sum_{\alpha \in S(p)} p(\alpha) \log(p(\alpha))$.

Definition 3. A WFSA $\mathcal{M}$ is defined by a square transition matrix M whose element m_{ij} give the probability of transition from state i to state j, an emission list h that gives a string $h_i \in \Sigma^*$ for each $i \neq 0$, and an acceptance vector $\vec{a}$ whose i-th component is 1 if i is an accepting state and 0 otherwise. There is a unique initial state which starts the state numbering at 0, and we permit states with empty outputs. Rows of M must sum to 1. Thus we have defined WFSA as normalized probability-weighted nondeterministic Moore machines.

Definition 4. The *weight* a WFSA assigns to a generation path is the product of the weights on the edges traversed, and the weight it assigns to a string α is the sum of the weights assigned to all paths that generate α.

1 Quantization error

The notions of quantization error and quantization noise, while well known in the signal processing literature (for a monographic treatment, see Widrow and Kollár 2008), and widely used in speech processing (Makhoul et al., 1985), have had little impact on language processing. Yet MDL description of even the simplest weighted language brings up a significant problem that cannot be addressed without approximation.

Let p be a non-computable real number between 0 and 1, and let us define the language A as containing only two strings, a and b, with probability p and $1 - p$ respectively. Since p is incompressible, the only way Alice can send A to Bob is by sending all bits of p. By Alice sending only the first n bits, Bob obtains a language A_n that approximates A with error of 2^{-n}. Since sending the strings $\{a, b\}$ has only a small constant cost, the overall MDL cost is dominated by the error term E, which is just as incompressible as the original p was.

As long as the weights themselves are treated as information objects of arbitrary capacity, there is no way out of this conundrum (de Leeuw 1956). On the other hand, the weighted languages we encounter in practice are generally abstracted from gigaword or smaller corpora, and as such their inherent precision is less than 32 bits. For weighted languages with finite support (corpora and language models without smoothing) p is simply a list containing strings and probabilities. The cost of transmitting this list comes from two sources: the cost of transmitting the probabilities, and the cost of transmitting the strings. As a first approximation, let us assume the two are independent, a matter we shall return to in Section 2.

We begin by investigating the inherent cost/error tradeoff of transmitting a discrete probability distribution $\{p_j | 1 \leq j \leq k\}$ by uniform quantization to b bits. We divide the unit interval in $n = 2^b$ equal parts. For our theorems we will use a value b large enough so that we have $p_j \geq 2^{-(b-2)}$ for all j, leaving at least the first 4 bins empty. Usually 32 bits suffice for this, and as we shall see shortly, often a lot fewer are truly needed, though standard modeling tools like SRILM often use 64-bit quantities. For each probability, Alice sends b bits (the bin number). Bob, who knows b, reconstructs a value based on the center of the bin.

Since this process does not guarantee that the reconstructed values sum to 1, Bob takes the additional step of renormalizing these values: if $\sum q_i = r$, he will use $\overline{q} = q_i/r$ instead of the q_i that were transmitted by Alice. When b is large, the p_i will be distributed uniformly mod 2^{-b}. In this case, the expected values $E(p_i - q_i)$ are zero for all i, so $E(\sum q_i) = \sum E(q_i) = \sum E(p_i) = E(\sum p_i) = 1$ or, in other words, $E(r) = 1$. Since $\text{Var}(r-1) = \sum_i \text{Var}(p_i - q_i) = k/12n^2$ is on the order $1/n^2$, in the following estimate we can safely ignore the effects of renormalization. By Definition 2, the KL approximation error is

$$Q = \sum_{i=0}^{n-1} \sum_{i/n \leq p_j \leq (i+1)/n} p_j \Delta(p_j^i) \qquad (1)$$

where $\Delta(p_j^i) = \log(2np_j/(2i+1))$ is the difference between the logarithms of the actual p_j and the centerpoint of the interval $[i/n, (i+1)/n)$ where p_j falls. In absolute value, this is maximal when p_j is at the lower end of this interval, where $\Delta(p_j^i)$ is $\log(\frac{2i+1}{2i})$. Using the standard estimate $\log(1+x) \leq x$ this will be less than $\frac{1}{2i} \leq 1/8$ since $i \geq 4$. Since the $\Delta(p_j^i)$ are now estimated uniformly, and the p_j sum to 1, we obtain

Theorem 1. The approximation error Q_n of uniform quantization into n bins $[i/n, (i+1)/n)$ such that the first 4 bins are empty satisfies

$$Q_n \leq \frac{1}{8 \log 2} \sim 0.18 \qquad (2)$$

bits independent of n (the computation was in base e rather than base 2, hence the factor $\log 2$). With growing n the number of bins that remain empty will grow, and the estimate $\frac{1}{2i}$ of Δ can be improved accordingly.

Theorem 1 of course gives just an upper bound, and a rather crude one, the expected value of Q_n is considerably less. Instead of using the max value $\Delta(p_j^i)$ we can consider the expected absolute value, which is $\log(1+\frac{1}{2i})/n$, so equation (2) could be reformulated as

$$E(Q_n) \leq \frac{1}{8n \log 2} \qquad (3)$$

It is evident from the foregoing that the crux of the matter are the small p_i values, and at any rate, there can only be a handful of relatively large values, since the sum is 1. Experience shows that probabilities obtained from corpora span many orders of magnitude, which justifies the use of a log scale. Instead of the simple uniform quantization of Theorem 1, we will use a two-parameter quantization scheme, whereby first $\log p_j$ are cut off at $-C$, and the rest, which are on the $(-C, 0)$ interval, are sorted in $n = 2^b$ bins 'b-bit quantization'.

In effect, all probabilities below e^{-C} are assumed to be below measurement precision, and the log of the rest are uniformly quantized. We experimented with two simple techniques: representing the class $(-\infty, -C)$ with a very low fixed value (10^{-50}) or with one set to e^{-2C} based on the parameter C of the encoding. As there was no appreciable difference (which is not surprising given $\lim_{x \to 0} x \log x = 0$), from here on we simply speak of *zero weights* for weights below e^{-C}.

We emphasize that 'being below measurement precision' is not the same as 'being zero' in the above sense. First, in any corpus of size N the smallest number we can measure is $1/N$, yet we know that further strings that were not in our sample are not necessarily probability zero. It is therefore common to reserve a small fraction,

	∅	a	akár	bár	egyvala	más	másvala	minden	se	vala
hány	72383	9502	2432	55					21	4584
hogy	7781539	213687	3173	1839		4570		123	4138	31873
hol	117231	399052	1037	9845		16066		16009	20521	34081
honnan	24777	18628	296	1205		2482		1321	627	4274
honnét	1598	1197	12	25		78		33	23	236
hová	17589	21073	486	1753	1	5073	1	1859	2249	3966
hova	17360	10591	309	1166		1788		1381	2105	3036
ki	1309618	1464744	3933	60923	884		814	308508	165230	221175
meddig	11879	8171	189	225					74	252
mely	761277	1586913	166	74262	3				4	40601
melyik	68051	47564	1996	34477	2				939	48274
mennyi	76429	25805	657	1415					517	96184
mi	1626013	1303820	6500	52480	1337		161		275773	355690
miért	251120	20672	58	205	4				1810	13552
mikor	173652	555325	679	33516		15892		11288	206	18235
milyen	343643	38921	8217	68033		1618	1		55603	81155

Table 1: Frequencies of proquants in the Hungarian Webcorpus

generally 1-5% of the probability mass, to unseen events, and use calculated numbers, instead of measured values, to smooth the distribution. Unfortunately, the engineering philosophy behind the various backoff schemes (which often utilize MDL stopping criteria both in speech and language modeling, see e.g. Shinoda and Watanabe 2000, Seymore and Rosenfeld 1996) is diametrically opposed to the the method of inquiry preferred by linguists, whose primary interest is with generalization, i.e. with models that make falsifiable predictions, rather than furnishing descriptive statistics. In particular, negative generalizations, that something is forbidden by some rule of grammar, are just as interesting from their standpoint as positive generalizations. But how do we express a negative generalization?

Definition 5. A string will be deemed *ungrammatical* or *structurally excluded* iff every generation path includes at least one zero weight in the above sense.

If scores from different sources are multiplied together, the use of zero weights as markers of ungrammaticality is implicit in the semantics of WFSA.[1] Still, there are significant difficulties in implementing the idea. If we want to maintain the commonsensical assumption that **teh* is not a word (has zero unigram weight) and also account for the data that makes it the 34,174th most common string in English text, we will need to model typos. Once we learn that the log price of the /the/teh/ substitution is about -9.8, we can predict not just the frequency of *teh,* but also those of *weatehr, otehr, tehy, tehre, tehft,* and so forth, without adding these to the lexicon. Since such a model is based on computed frequencies of letter substitution and exchange rather than on the typos directly, the engineer has to give up the enterprise of building the entire language model in a single sweep directly on the data.

At the same time, the linguist has to give up the attractive simplicity of 'zero weight iff ungrammatical': the misspelling model will assign a low but nonzero weight to everything, and if this model is compiled together with a unigram model that contains only grammatical words, the simple world-view of Definition 5 will no longer work. Rather, we will have to say that it is zero weight in the *grammatical* subautomaton (visible only prior to getting compiled together with the semantic, spelling, stylistic, and possibly other subautomata) that defines grammaticality. We have to build an explicit noise model to make sense of the raw data, but this is not particularly surprising from the perspective of other sciences like astronomy where noise reduction is common practice.

The specific contribution of the MDL approach is that zero weights in the model are *a lot* cheaper than using low probabilities would be: the paradigm encourages both sparseness and structural decomposition. But before we can establish

[1]We owe this observation to an anonymous MOL referee.

these points in Sections 2 and 3, we need to assimilate another piece of computational practice, the use of log probabilities. When quantization is uniform on the log scale, the expected value of the binning error is no longer zero, given our assumption of uniformity on the linear scale, but rather

$$\int_{-C\log\frac{i+1}{n}}^{-C\log\frac{i}{n}} e^x - e^{-C\frac{i+0.5}{n}} dx \sim C^3/8n^3 \qquad (4)$$

which yields an expected $r \sim kC^3/8n^3$, still negligible compared to the bound given in Theorem 2, which is obtained by methods similar to those used above.

Theorem 2. For C, n sufficiently large for the first 4 bins to remain empty, the approximation error L_n^C of log-uniform quantization with cutoff $-C$ into $n = 2^b$ bins $[-C(i+1)/n, -Ci/n)$ is bounded by

$$L_n^C \leq \frac{C}{2n\log 2} \qquad (5)$$

and the expected value $E(L_n^C)$ is bounded by $C^2/4n^2\log 2$.

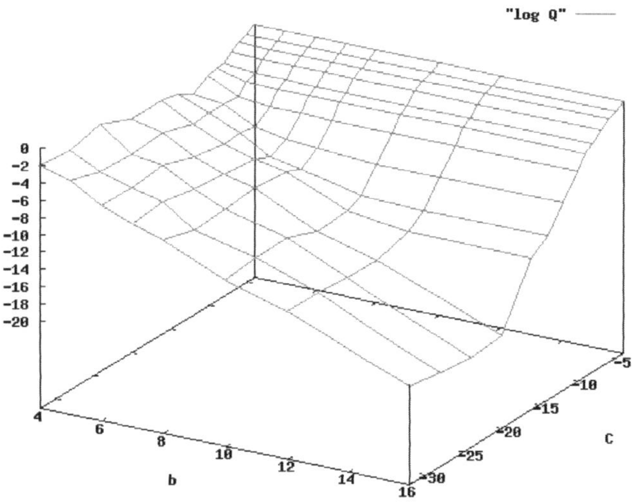

Figure 1: Error of log-scale uniform quantization

Let us see on an example how these error bounds compare to values obtained numerically. Our first example will explore what we will call, for want of a better name, the proquant system of Hungarian that covers both pro-forms (pronouns, proadjectives, proadverbials) and quantifiers. Given the prefixes *a-*, *minden-*, *vala-*, *egyvala-*, *másvala-*, *se-*, *akár-*, *más-*, *bár-* and zero, and suffixes *-ki*, *-mi*, *-hol*, *-hogy*, *-hova*, *etc.* we can create forms such as *valaki* 'someone', *valami* 'something', *akárki* 'anyone', *sehol* 'nowhere' and so on. Clearly, many of what we call prefixes and suffixes could be analyzed further, e.g. *másvala* as *más+vala*, but we don't want to prejudge the issue by presenting a maximally detailed analysis.

In a corpus of over 40 million sentences (Hungarian Webcorpus, Halácsy et al. 2004) we observed the frequencies in Table 1. Many of these proquant forms take inflectional suffixes (case, number, etc.), and the numbers presented here already include these, so that the 814 occurrences of *másvalaki* include forms like *másvalakivel* 'with someone else', *másvalakinek* 'to someone else' etc. If we think of the (stemmed) Hungarian vocabulary as a weighted language h, the set of prefixes (suffixes) as an unweighted language Pre (resp. Suff), the data is a sample from $S(h)\cap$ Pre·Suff with the weights renormalized. Altogether, we have 121 nonzero values plus 39 zeros, the entropy of the distribution is $H = 3.677$. Figure 1 plots the log of the observed quantization noise as a function of the number of bits b and the cutoff $-C$. Notice that once C is sufficiently large, no further gains are made by increasing it further. As expected from Theorem 2, the log of the error is roughly linear in $b = \log_2 n$ (the observed values are of course better than the bounds).

Definition 6. The *inherent noise* of a dataset D is the KL approximation error between a random subsample and its complement.

Ideally, we would want to compare another sample D' to D, but in many cases launching a comparable data collection effort is simply not feasible, and we must content ourselves with the simple procedure suggested by this definition. By randomly cutting the 40m sentence corpus on which the proquant dataset is based in 10m sentence parts and computing the KL divergence between any two, we obtain numbers in the 7-8·10^{-5} range, which means it makes little sense to approximate D with better precision than 10^{-5}. How to handle the singular cases when some q_j becomes 0 (as happens with half of the hapaxes when we cut the sample in two) is an issue we defer to Section 3.

Since the smallest p_j in this data is about $5{\cdot}10^{-8}$, by taking $C = 20$ we guarantee that no log probability is less than the cutoff point $-C$. Trivial 'list' automata consisting of an initial state, a final state, and a separate Mealy arc (or Moore state) for each of the 121 nonzero observations already generate a weighted language within the inherent noise of the data at 10 bits, where the KL divergence is at $8 \cdot 10^{-6}$. At 12 bits, the divergence is below $1.4 \cdot 10^{-6}$, and at 16 bits, below $5 \cdot 10^{-9}$. As we shall see in the next Section, the MDL size of these models, between 2k and 7k bits, is dominated by factors unrelated to the precision b of the encoding.

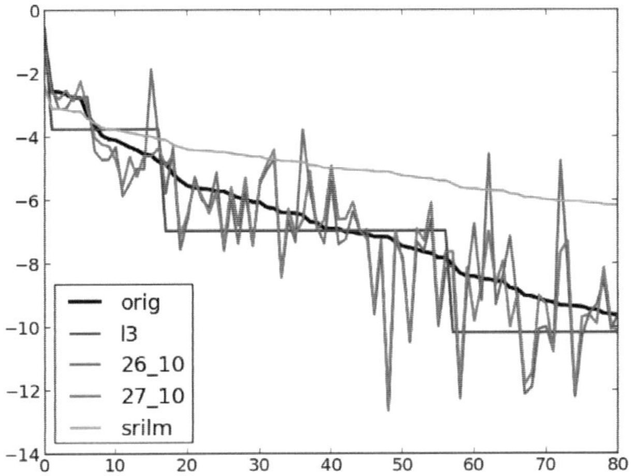

Figure 2: Model fit to observed probabilities

Figure 2 shows the 80 largest observed proquant probabilities (in black) in descending order, and the probabilities of the same strings as computed from several models. The 10 and 12-bit list automata are not plotted, as the computed values are graphically indistinguishable from the observed values, the rest will be discussed in the text.

2 Detecting ambiguity

Before turning to the actual MDL learning process, let us summarize what we have for the Hungarian proquant system so far. We have a weighted language of about 120 strings. When transmitting a weighted automaton, Alice is sending not strings and weights, but rather weight-labeled arcs and string-labeled states of a WFSA. In Definition 3 we used Moore machines, but in the literature Mealy automata, where inputs/outputs and weights are both tied to arcs are more common (see e.g. Mohri 2009). The rationale for preferring Moore over Mealy in the MDL context is that no gains can be obtained from joint compression of strings and probabilities (even though Mealy machines couple the two), while sharing of strings has very significant impact on MDL length, as we shall see shortly. For the simple 'list' automata this means adding extra states in the middle of a Mealy arc, and we need to take some precautions to guarantee that the representation is just as compact as it would be for a Mealy machine.

Let us now see in some detail how compact these encodings can get. With s states, and b bits for probability, an arc requires $2\log_2 s + b$ bits. However, Bob can reasonably assume that Alice is only sending trimmed machines, with states that cannot be reached from the initial state or with no path to an accepting state already removed. Therefore, if Bob sees a state with no outbound path he supplies an outgoing arc, with probability 1, to the final state – such arcs need not be sent by Alice to begin with. Similarly, Bob can assume that all states except for the last one are non-accepting, and Alice will transmit information only to override this default when needed.

As for emissions, in a Moore machine each state emits a string (but no guarantees that different states emit different strings), so Alice needs to encode the strings somehow. If we assume that there is a character table shared between Alice and Bob, e.g. the character frequencies of Hungarian, with entropy H, encoding a string α costs simply $|\alpha|H$ bits. (We could take this also to be a case of transmitting a weighted language, but we assume that the cost of transmitting this language can be amortized over many WFSA that deal with Hungarian.)

b	l	M	c_s	c_a	KL	H_q
1	121	5210	4306	904	2.1883	6.833
2	121	5386	4306	1080	1.1207	3.487
3	121	5507	4306	1201	0.268	2.889
4	121	5628	4306	1322	0.041436	4.044
5	121	5749	4306	1443	0.016117	3.424
6	121	5870	4306	1564	0.002409	3.667
7	121	5991	4306	1685	0.000676	3.653
8	121	6112	4306	1806	0.000288	3.647
9	121	6233	4306	1927	5.905e-5	3.681
10	121	6354	4306	2048	8.003e-6	3.678
11	121	6475	4306	2169	3.999e-6	3.678
12	121	6596	4306	2290	1.387e-6	3.678
16	121	7080	4306	2774	4.660e-9	3.676

Table 2: List models with character-based string encoding

Table 2 summarizes the relevant values for the trivial models where each weight gets its own trainable parameter. b is the number of bits, l is the number of trainable parameters (weights associated to arcs), c_a is the cost of transmitting the arcs. Note that this is less than $l(b+2\log_2 s)$, because of the redundancy assumptions shared by Alice and Bob. c_s is the cost of transmitting the emissions, and the total model cost is $M = c_a + c_s$. We would, ideally, also need to add to M a dozen bits or so to encode the major parameters of the coding scheme itself, such as the values $b = 10$ and $C = 20$, but these turn out to be negligible compared to the basic cost. Also, these major parameters are shared across the alternatives we compare, so whatever we do to minimize M will not be affected by uniformly adding (or uniformly ignoring) this constant cost. KL gives the KL divergence between the model and the training data. This measures the expected extra message length per arc weight, so that the error residual E is k times this value, where k is the number of values being modeled. We emphasize that $k = l$ only in the listing format, where all values are treated as independent – in the 'hub' model we shall discuss shortly l is only 26 (10 prefix and 16 suffix weights) but k is still 121.

The main components of the total MDL cost, $M, l \cdot KL, l(KL + H_q)$, and the total $M + l(KL + H_q)$ are plotted on Figure 3.

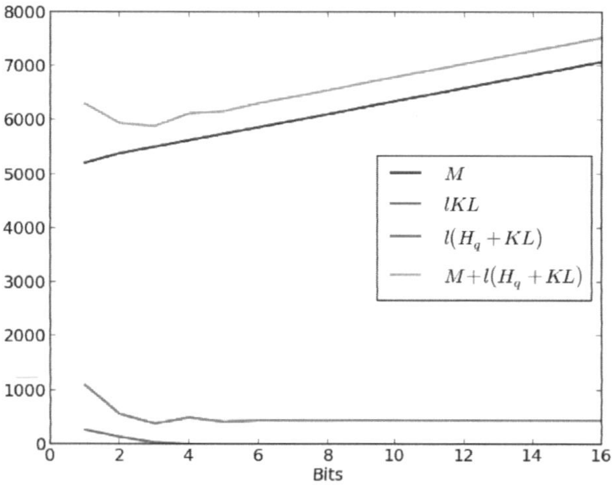

Figure 3: MDL cost components

All models with $b \geq 10$ are within the internal noise of the data, and it takes over 6kb to describe such a model. However, the bulk of these bits come from encoding the output strings character by character – if we assume that Alice and Bob share a morpheme table, the results improve a great deal, by over 3,600 bits. If the system recognizes what we already anticipated in Table 1, that each string can be expressed as the concatenation of a prefix and suffix, encoding the strings becomes drastically cheaper. Using MDL for segmentation is a well-explored area (see in particular Goldsmith 2001, Creutz and Lagus 2002, 2005), and we are satisfied by pointing out that using the morphemes in the first row and column of Table 1 we drastically reduce c_s, to about 708 bits, below the cost c_a of encoding the probabilities. The 3-bit list model providing the MDL optimum (dark blue in Figure 2) requires 1,900 bits with this string encoding, and is noticeably better than the SRILM bigram/trigram (turquoise) which takes around 12kb.

By encoding the emissions in a more clever fashion, we have not changed the structure of the model: the same states are still linked by the same arcs carrying the same probabilities, it is just the state labels that are now encoded differently. When expressed as a Mealy automaton, a listing of probabilities corresponds to a two-state WFSA with as many arcs as we have list elements (in our case, 121), while the arrangement of Table 1 is suggestive of a different model, one with 10 prefix arcs from the initial state to a central 'hub' state, and 16 suffix arcs from this hub to the final state.

We have trained such 'hub' models using KL, Euclidean (L_2), and absolute value (L_1) minimization techniques. Of these, direct minimization of KL divergence works best, obtaining 0.325 bits at $b = 10$, and 0.298 at $b = 12$ (red and green in Figure 2). While the difference, about 0.027 bits, is still perceptible compared to the noise level, with a signal to noise ratio (SNR) of 8 dB, it simply does not amortize over the 26 model probabilities we need to encode. Adding 2 bits for encoding one value requires a total of adding 52 bits to our specification of $\mathcal{M}$, while the gain of the error residual E, computed over the 121 observed values, is just 2.074 bits. In short, there is not much to be gained by going from 10 to 12 bits, and we need to look elsewhere for further compression.

Definition 7. For a weighted language p a model transform X is *learnable in principle* (LIP) if (i) both $\mathcal{M}$ and $X(\mathcal{M})$ are part of the hypothesis space and (ii) the total MDL cost of describing p by $X(\mathcal{M})$ is significantly below that of describing p by $\mathcal{M}$.

In a critical sense, LIP is weaker than MDL learnability, since the space itself can be very large, and testing all hypothetical transforms X that fit the bill may not be feasible. The difference between LIP and practical MDL learnability is precisely the difference between existence proofs and constructive proofs. Our interest here is with the former: our goal is to demonstrate that structurally sound models are LIP. So far, we have seen that structurally valid segmentations can be effectively obtained by MDL. Our next task is to show that *ambiguity* is LIP.

As linguists, we know that the weakest point of the hub model is that *hogy*, accounting for almost 40% of the data, is not just a proquant 'how' but also a subordinating conjunction 'that'. To encode this ambiguity, we add another arc emitting *hogy* directly. Table 3 compares list models (lines 1-3, emissions encoded over morphemes rather than characters), simple hub models (lines 4-6), and hub models with this extra arc (lines 7-9).

b	l	M	c_s	c_a	KLe5	$M+E$
3	121	1907	705	1202	26800	2289
10	121	2754	705	2049	0.8	3199
12	121	2999	705	2290	0.14	3441
3	26	473	81	392	42343	1305
10	26	662	81	581	32593	1201
12	26	716	81	635	29827	1249
3	27	480	81	400	23094	1052
10	27	676	81	596	11268	1161
12	27	733	81	652	10022	1198

Table 3: Hub models with/out ambiguous *hogy*

As can be seen from the table, the best model again takes only 3 bits, but must include the extra parameter for handling the ambiguity of *hogy*. To learn this, at least in principle, without relying on the human knowledge that drove the heuristic search, consider the leading terms of the KL error. Arranging the $p_i \log(p_i/q_i)$ in order of decreasing absolute value we obtain *mi* 0.0192; *minden+ki* 0.0175; *a+mely* 0.0169; *mely* -0.0147; *a+mikor* 0.0135; *hogy* -0.0128; and so forth. Of all the 121 strings we may consider for direct emission, only *hogy* is worth adding a separate arc for. Further, if we repeat the process, adding a second direct arc never results in sufficient entropy gain compared to adding *hogy* alone.

To summarize, list models can approximate the original data within its inherent noise level, but incur a very significant MDL cost, even if they use an efficient string encoding because they keep many parameters, see the first three lines of Table 3 above. The hub models, which build structure similar to the one used in the string encoding, recognizing prefixes and suffixes for what they are, are far more compact, at 470-730 bits, even though they have a KL error of about .1-.4 bits. Finally, the hub+ambiguity model, with 27 parameters, reduces the total MDL cost to 1052 bits, less than half of the best list model.

Currently we lack the kind of detailed understanding of the description length surface over the WFSA$\times$stringencoding space that would let us say with absolute certainty that e.g. the hub model with ambiguous *hogy* is the global minimum, and we cannot muster the requisite computational power to exhaustively search the space of all WFSA with 27 arcs or less. Further gains could quite possibly made with even cruder quantization, e.g. to $n = 6$ levels (powers of 2 are convenient, but not essential for the model), or by bringing in non-uniform quantization.

On the one hand, we are virtually certain that the only encoding of emissions worth studying is the morpheme-based one, since the economy brought by this is tremendous, 3,600 bits over the proquants alone, and no doubt further gains elsewhere, as we extend the scope to other words that contain the same morphemes – in this regard, our findings simply confirm what Goldsmith, Creutz, Lagus, and others have already demonstrated. On the other hand, finding the right segmentation is only the first step, we also need a good model of the tactics. As we said at the beginning, the encoding of arcs and probabilities can to a significant extent be independent of the encoding of the emissions. Here the remarkable fact is that a better emission model could to a large extent drive the search for structuring the WFSA itself.

Given a segmentation of a string $\alpha = \alpha_1\alpha_2$, the hypothesis space includes both a single arc from some r to some t where we emit α, or the concatenation of two arcs $r \rightarrow s$ and $s \rightarrow t$ with s and t emitting α_1 and α_2 respectively. This brings in a bit of ambiguity in regards to the distribution of the probabilities, for if α had weight p the new arcs could be assigned any values p_1, p_2 as long as $p_1p_2 = p$, at least if the sum of outgoing probabilities from s remains 1. If s has no other arcs outgoing than $s \rightarrow t$ this forces $p_1 = p$, but if we collapse the intermediate states from several bimorphemic words, there is room for joint optimization. In our example, collapsing all intermediate states in a single 'hub' halves the MDL cost.

3 Decomposition

For our next example we consider Hungarian stem-internal morphotactics. The Analytic Dictionary of Hungarian (Kiss et al 2011) provides, for each stem like *beleilleszt* 'fit in' an analysis like preverb+root+suffix wherein *bele* is one of a closed set of Hungarian preverbal particles, *ill* is the root, and *eszt* is a verb-forming suffix. There are six analytic categories: Stem S; sUffix U; Preverb P; root E; Modified M; and foreIgn I; so that each stem gets mapped on a string over $\Sigma = \{S, U, P, E, M, I\}$. We have two weighted languages: the *tYpe-weighted* language Y where each string is counted as many times as there are word types corresponding to it (so that e.g. for SUU we have 3,739 stems from *ábrándozik* 'daydream' to *zuhanyozó* 'shower stall', and the *tOken-weighted* language O where the same pattern has weight 18,739,068 because these words together appeared that many times in the Hungarian Webcorpus (Halácsy et al., 2004).

Since the inherent noise of O is about 0.0474 bits, we are interested in automata that approximate it within this limit. This is easily achieved with HMM-like WFSA that have arcs between any two states, using $b = 11$ bits or more, the smallest requiring only 781 bits. For Y the inherent noise is less, 0.011 bits, and the complete graph architecture, which only has 49 parameters (6 states, plus arcs from an initial state and arcs to a final state) is not capable of getting this close to the data, with the best models, from $b = 11$ onwards, remaining at KL distance 0.3. The two languages differ quite markedly in other respects as well, as can be seen from the fact that the character entropy of O is 0.933, that of Y is 1.567. Type frequency is not a good predictor of token frequency: the KL approximation error of O relative to Y is 2.11 bits.

An important aspect of the MDL calculus is the treatment of the singularities which arise whenever some of the q_i in Definition 2 are 0. In the case at hand, we find both types that are not attested in the corpus, and tokens whose type was not listed in the Analytic Dictionary, a situation that would theoretically render it impossible to compute the KL divergence in either direction. In practice, tokens with no dictionary type are either collected in a single 'unknown' type or are silently discarded. Both techniques have merit. The catch-all 'unknown' type can simply be assumed to follow the distribution of the known types, so a model that captures the data enshrined in the dictionary should, at least in principle, be also ideal for the data not seen by the lexicographer. Surprises may of course lurk in the unseen data, but as long as coverage is high, say $P(\text{unseen}) \leq 0.05$, surprises will really be restricted to this 5% of the unseen, or what is the same, will be at order P^2. In general we may consider two distributions $\{p_i\}$ and $\{q_i\}$ as in Definition 2, and compute $P = \sum_{q_i=0} p_i$, the proportion of q-singular data in p.

Theorem 3. The total cost L of transmitting an item from the p-distribution is bound by

$$L \leq (1-P)(KL(p,q)+H_q)+P(1+\log_2 n) \quad (6)$$

Proof We use, with probability $(1 - P)$, the q-based codebook: this will have cost H_q plus the modeling loss $KL(p,q)$. In the remaining cases (probability P) we should use a codebook based on the q-singular portion of p, but we resort to uniform coding at cost $\log_2 n$, where n is the number of singular cases. We need to transmit some information as to which codebook is used: this requires an extra $H(P, 1-P) \leq P$ bits – collecting these terms gives (6). □

Theorem 3 gives a principled basis for discarding data within the MDL framework: when P is small, the second term of (6) can be absorbed in the noise. To give an example, consider the unigram frequencies listed in columns f_O and f_Y of Table 4. Some letters are quite rare, in particular, I makes up less than 0.06% of O and 0.013% of Y. Columns KL_O and KL_Y show the KL divergence of O and Y from models obtained by by discarding words containing the letter in question, columns P_O and P_Y show the weight of the strings that are getting discarded.

	f_O	P_O	KL_O	f_Y	P_Y	KL_Y
S	.7967	.9638	4.7887	.5342	.9122	3.5092
U	.1638	.1464	0.2284	.3443	.5699	1.2174
M	.0083	.0103	0.0149	.0255	.0623	0.0928
E	.0114	.0141	0.0205	.0331	.0804	0.1209
P	.0198	.0248	0.0362	.0623	.1531	0.2397
I	.0001	.0001	0.0002	.0006	.0010	0.0006

Table 4: Divergence caused by discarding data

In the case of Y, only I can be discarded while keeping below the inherent noise of the data, but for O we have three other symbols *M, E,* and *P*, that could be removed. Further, removing both letters M, E only produces a KL loss of 0.036 bits; removing M, I a loss of 0.015 bits; E, I 0.021

bits; P, I 0.036 bits; and even removing all three of M, E, I only 0.036 bits.

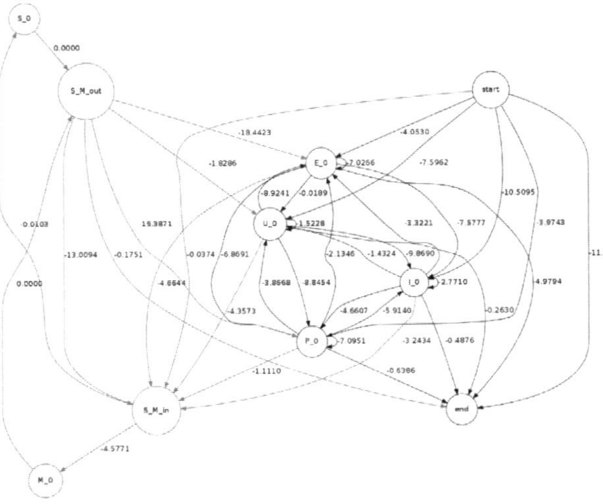

Figure 4: MS merge-split model

In the case of I, again as linguists we understand quite well what discarding this data means: we are excluding foreign stems. This is quite justified, not because foreign words like *paperback, pacemaker* or *baseball* are in any way inferior, but because their internal analysis is not transparent to the Hungarian reader (it is telling that the editors of the Analytic Dictionary coded the stem boundary in *paper+back* but not in *base+ball*).

Discarding M, a category that differs from S only in that the stem undergoes some automatic morphophonological change such as vowel elision, is also a sensible step in that the fundamental morphotactics are not at all affected by these changes, but how is this learnable, even in principle? Here we introduce another model transform called *XY merge-split* composed of two steps: first we replace all letters (or strings) X by Y and train a model, and next we split up the emission states of Y in the merged model to X and Y-emissions according to the relative proportions of X and Y in the original data.

For LIP, the key observation is that models constructed by *XY merge-split* have a transmission cost composed of two parts, the length of the smaller merged model (given in black in Figure 2), plus transmitting the pair X, Y and the probability of the split, which is exactly the cost of a single arc, even though the actual split model will have many more arcs (given in red in Figure 4). Once this is taken into account, we can systematically investigate all 6·5 merge-split possibilities. The results confirm the educated linguistic guess quite remarkably. The best compression rates are obtained by merging I with any of the minor categories or, if I is already discarded or merged in, merging M into S. The smallest O model before these steps took 781 bits, this is now reduced to 502 bits. If we start by discarding I, and merging M to S afterwards, this can be reduced to 349 bits. In the end we merge the morphophonologically affected forms with the ones not so affected not because our training as linguists tells us we should do this, but because that is what brevity demands.

4 Conclusions

In this paper we have developed an MDL-based framework for structure detection based on simple notions mostly borrowed from signal processing: quantization noise, inherent noise level, and cutoffs. Standard n-gram models fare rather poorly compared either in size or in model accuracy to the WFSA results obtained here: for example on the morphotactics data a straight SRILM trigram model has over 200 parameters and has KL divergence 1.09 bits. Most of the 64 bits per n-gram parameter are wasted (if we assume only 12 bits per parameter, the WFSA we use requires only 49 parameters and gets within 0.03 bits of the observed data) and further, the general-purpose backoff scheme built into SRILM just makes matters worse.

Similarly, on the proquant data an SRILM bigram model has 175 parameters (including the 26 unigram weights but excluding the backoff weights), yet it is farther from the data at 64 bits resolution than our best 27-parameter model at 3 bits. More important, the bigram structure of the proquant data has to be hand-fed into the standard model, while the MDL approach can discover this, together with other linguistically relevant observations such that *hogy* was ambiguous.

This is not to say that n-gram models are no longer competitive, for our current MDL methods, based on a simulated annealing learner, use too much CPU and will not scale to the gigaword regime without much further work. Yet if formal grammar and information theory are to get together again, as (Pereira, 2000) suggests, we must direct effort towards recapitulating linguistic practice, including the 'dirty' parts such as discarding data strategically. The main thrust of the work presented here is that the data manipulation methods that are the stock in trade of the descriptive linguist are LIP, and Universal Grammar is simply a short list of the permissible model transformations including path duplication for ambiguity, state merging for position class effects, and merge-split for collapsing categories.

Acknowledgments

We thank Dániel Varga (Prezi) and Viktor Nagy (Prezi) for the first version of the simulated annealing WFSA learner. Zséder wrote the version used in this study, Recski collected the data and ran the HMM baseline, Kornai advised. The current version benefited greatly from the remarks of anonymous referees. Work supported by OTKA grants #77476 and #82333.

References

Noam Chomsky and Morris Halle. 1965a. Some controversial questions in phonological theory. *Journal of Linguistics*, 1:97–138.

Kenneth Church, Ted Hart, and Jianfeng Gao. 2007. Compressing trigram language models with Golomb coding. In *Proceedings of the Joint Conference on Empirical Methods in Natural Language Processing and Computational Natural Language Learning*, pages 199–207, Prague. ACL.

Mathias Creutz and Krista Lagus. 2002. Unsupervised discovery of morphemes. In *Proc. 6th SIGPHON*, pages 21–30.

Mathias Creutz and Krista Lagus. 2005. Unsupervised morpheme segmentation and morphology induction from text corpora using Morfessor 1.0. Technical Report A81, Helsinki University of Technology.

Karel de Leeuw, Edward F. Moore, Claude E. Shannon, and N. Shapiro. 1956. Computability by probabilistic machines. In C.E. Shannon and J. McCarthy, editors, *Automata studies*, pages 185–212. Princeton University Press.

Samuel Eilenberg. 1974. *Automata, Languages, and Machines*, volume A. Academic Press.

John A. Goldsmith. 2001. Unsupervised learning of the morphology of a natural language. *Computational Linguistics*, 27(2):153–198.

Peter Grünwald. 1996. A minimum description length approach to grammar inference. In Stefan Wermter, Ellen Riloff, and Gabriele Scheler, editors, *Conectionist, statistical, and symbolic approaches to learning for natural language processing*, LNCS 1040, pages 203–216. Springer.

Péter Halácsy, András Kornai, László Németh, András Rung, István Szakadát, and Viktor Trón. 2004. Creating open language resources for Hungarian. In *Proceedings of the 4th international conference on Language Resources and Evaluation (LREC2004)*, pages 203–210.

Frederick Jelinek. 1997. *Statistical Methods for Speech Recognition*. MIT Press.

Gábor Kiss, Márton Kiss, Balázs Sáfrány-Kovalik, and Dorottya Tóth. 2011. A magyar szóelemtár megalkotása és a magyar gyökszótár előkészítő munkálatai. In A. Tanács and V. Vincze, editors, *MSZNY 2012*, pages 102 – 112.

John Makhoul, Salim Roucos, and Herbert Gish. 1985. Vector quantization in speech coding. *Proceedings of the IEEE*, 73(11):1551–1588.

Mehryar Mohri. 2009. Weighted automata algorithms. In Manfred Droste, Werner Kuich, and Heiko Vogler, editors, *Handbook of Weighted Automata*, Monographs in Theoretical Computer Science, pages 213–254. Springer.

Fernando Pereira. 2000. Formal grammar and information theory: Together again? *Philosophical Transactions of the Royal Society*, A 358:1239–1253.

Jorma Rissanen. 1978. Modeling by the shortest data description. *Automatica*, 14:465–471.

Arto Salomaa and Matti Soittola. 1978. *Automata-Theoretic Aspects of Formal Power Series*. Springer, Texts and Monographs in Computer Science.

Kristie Seymore and Ronald Rosenfeld. 1996. Scalable backoff language models. In *Spoken Language, 1996. ICSLP 96. Proceedings., Fourth International Conference on*, volume 1, pages 232–235. IEEE.

Koichi Shinoda and Takao Watanabe. 2000. MDL-based context-dependent subword modeling for speech recognition. *Journal of the Acoustical Society of Japan (Eenglish edition)*, 21(2):79–86.

Ray J. Solomonoff. 1964. A formal theory of inductive inference. *Information and Control*, 7:1–22, 224–254.

Paul M. B. Vitanyi and Ming Li. 2000. Minimum description length induction, Bayesianism, and Kolmogorov complexity. *IEEE Transactions on Information Theory*, 46(2):446–464.

Bernard Widrow and István Kollár. 2008. *Quantization Noise: Roundoff Error in Digital Computation, Signal Processing, Control, and Communications*. Cambridge University Press.

Why Letter Substitution Puzzles are Not Hard to Solve: A Case Study in Entropy and Probabilistic Search-Complexity

Eric Corlett
University of Toronto
10 King's College Rd., Room 3302
Toronto, ON, Canada M5S 3G4
ecorlett@cs.toronto.edu

Gerald Penn
University of Toronto
10 King's College Rd., Room 3302
Toronto, ON, Canada M5S 3G4
gpenn@cs.toronto.edu

Abstract

In this paper we investigate the theoretical causes of the disparity between the theoretical and practical running times for the A^* algorithm proposed in Corlett and Penn (2010) for deciphering letter-substitution ciphers. We argue that the difference seen is due to the relatively low entropies of the probability distributions of character transitions seen in natural language, and we develop a principled way of incorporating entropy into our complexity analysis. Specifically, we find that the low entropy of natural languages can allow us, with high probability, to bound the depth of the heuristic values expanded in the search. This leads to a novel probabilistic bound on search depth in these tasks.

1 Introduction

When working in NLP, we can find ourselves using algorithms whose worst-case running time bounds do not accurately describe their empirically determined running times. Specifically, we can often find that the algorithms that we are using can be made to run efficiently on real-world instances of their problems despite having theoretically high running times. Thus, we have an apparent disparity between the theoretical and practical running times of these algorithms, and so we must ask why these algorithms can provide results in a reasonable time frame. We must also ask to what extent we can expect our algorithms to remain practical as we change the downstream domains from which we draw problem instances.

At a high level, the reason such algorithms can work well in the real world is that the real world applications from which we draw our inputs do not tend to include the high complexity inputs. In other words, our problem space either does not cover all possible inputs to the algorithm, or it does, but with a probability distribution that gives a vanishingly small likelihood to the "hard" inputs. Thus, it would be beneficial to incorporate into our running time analysis the fact that our possible inputs are restricted, even if only restricted in relative frequency rather than in absolute terms.

This means that any running time that we observe must be considered to be dependent on the distribution of inputs that we expect to sample from. It probably does not come as a surprise that any empirical analysis of running time carries with it the assumption that the data on which the tests were run are typical of the data which we expect to see in practice. Yet the received wisdom on the asymptotic complexity of algorithms in computational linguistics (generally what one might see in an advanced undergraduate algorithms curriculum) has been content to consider input only in terms of its size or length, and not the distribution from which it was sampled. Indeed, many algorithms in NLP actually take entire distributions as input, such as language models. Without a more mature theoretical understanding of time complexity, it is not clear exactly what any empirical running time results would mean. A worst-case complexity result gives a guarantee that an algorithm will take no more than a certain number of steps to complete. An average-case result gives the expected number of steps to complete. But an empirical running time found by sampling from a distribution that is potentially different from what the algorithm was designed for is only a lesson in how truly different the distribution is.

It is also common for the theoretical study of asymptotic time complexity in NLP to focus on the worst-case complexity of a problem or algorithm rather than an expected complexity, in spite of the existence for now over 20 years of methods for average-case analysis of an algorithm. Even these, however, often assume a uniform distribution over input, when in fact the true expectation must consider the probability distribution that we will draw the inputs from. Uniform distributions are only common because we may not know what the distribution is beforehand.

Ideally, we should want to characterize the running time of an algorithm using some known properties of its input distribution, even if the precise distribution is not known. Previous work that attempts this does exist. In particular, there is a variant of analysis referred to as *smoothed analysis* which gives a bound on the average-case running time of an algorithm under the assumption that all inputs are sampled with Gaussian measurement error. As we will argue in Section 2, however, this approach is of limited use to us.

We instead approach the disparity of theoretical and practical running time by making use of statistics such as entropy, which are taken from the input probability distributions, as eligible factors in our analysis of the running time complexity. This is a reasonable approach to the problem, in view of the numerous entropic studies of word and character distributions dating back to Shannon.

Specifically, we analyze the running time of the A^* search algorithm described in Corlett and Penn (2010). This algorithm deciphers text that has been enciphered using a consistent letter substitution, and its running time is linear in the length of the text being deciphered, but theoretically exponential in the size of the input and output alphabets. This naïve theoretical analysis assumes that characters are uniformly distributed, however. A far more informative bound is attainable by making reference to the entropy of the input. Because the algorithm takes a language model as one of its inputs (the algorithm is guaranteed to find the model-optimal letter substitution over a given text), there are actually two input distributions: the distribution assumed by the input language model, and the distribution from which the text to be deciphered was sampled. Another way to view this problem is as a search for a permutation of letters as the outcomes of one distribution such that the two distributions are maximally similar. So our informative bound is attained through reference to the cross-entropy of these two distributions.

We first formalize our innate assumption that these two distributions are similar, and build an upper bound for the algorithm's complexity that incorporates the cross-entropy between the two distributions. The analysis concludes that, rather than being exponential in the length of the input or in the size of the alphabets, it is merely exponential in the cross-entropy of these two distributions, thus exposing the importance of their similarity. Essentially, our bound acts as a probability distribution over the necessary search depth.

2 Related Work

The closest previous work to the analysis presented here is the use of smoothed analysis to explain the tractable real-world running time of a number of algorithms with an exponential worst-case complexity. These algorithms include the simplex algorithm, as described by Spielman and Teng (2004), the k-means clustering algorithm, as described by Arthur et al. (2009) and others. As in our current approach, smoothed analysis works by running a general average-case analysis of the algorithms without direct knowledge of the distribution from which the problem inputs have been drawn. The assumption made in smoothed analysis is that every input has been read with some Gaussian measurement error. That is, in a typical worst-case analysis, we may have an adversary choose any input for our algorithm, after which we must calculate how bad the resulting running time might be, but in a smoothed analysis, the adversary gives us input by placing it into the real world so that we may measure it, and this measurement adds a small error drawn from a Gaussian distribution to the problem instance. The point of smoothed analysis is to find the worst average-case running time, under these conditions, that the adversary can subject us to. Thus the analysis is an average case, subject to this error, of worst cases. In the papers cited above, this method of analysis was able to drop running times from exponential to polynomial.

It is unfortunate that this approach does not readily apply to many of the algorithms that we use in NLP. To see why this is, simply note that we can only add a small Gaussian error to our inputs if our inputs themselves are numerical. If the inputs to our algorithms are discrete, say, in the form of strings, then Gaussian errors are not meaningful. Rather, we must ask what sort of error we can expect to see in our inputs, and to what extent these errors contribute to the running time of our algorithms. In the case of decipherment, "error" is committed by substituting one character for another consistently.

The strongest known result on the search complexity of A^* is given in Pearl (1984). This work found that, under certain assumptions, a bound on the absolute error between the heuristic used and the true best cost to reach the goal yields a polynomial worst-case depth for the search. This happens when the bound is constant across search instances of different sizes. On the other hand, if the relative error does not have this constant bound, the search complexity can still be exponential. This analysis assumes that the relative errors in the heuristic are independent between nodes of the search tree. It is also often very difficult even to calculate the value of a heuristic that possesses such a bound, as it might involve calculating the true best cost, which can be as difficult as completely solving a search problem instance (Korf et al., 2001). Thus, most practical heuristics still give rise to theoretically exponential search complexities in this view.

In Korf and Reid (1998) and Korf et al. (2001), on the other hand, several practical problems are treated, such as random k-SAT, Rubik's cubes, or sliding tile puzzles, which are not wholly unlike deciphering letter substitution puzzles in that they calculate permutations, and therefore can assume, as we do, that overall time complexity directly corresponds to the number of nodes visited at different depths in the search tree that have a heuristic low enough to guarantee node expansion. But their analysis assumes that it is possible to both estimate and use a probability distribution of heuristic values on different nodes of the search graph, whereas in our task, this distribution is very difficult to sample because almost every node in the search graph has a worse heuristic score than the goal does, and would therefore never be expanded. Without an accurate idea of what the distribution of the heuristic is, we cannot accurately estimate the complexity of the algorithm. On the other hand, their analysis makes no use of any estimates of the cost of reaching the goal, because the practical problems that they consider do not allow for particularly accurate estimates. In our treatment, we find that the cost to reach the goal can be estimated with high probability, and that this estimate is much less than the cost of most nodes in the search graph. These different characteristics allow us to formulate a different sort of bound on the search complexity for the decipherment problem.

3 The Algorithm

We now turn to the algorithm given in Corlett and Penn (2010) which we will investigate, and we explain the model we use to find our bound.

The purpose of the algorithm is to allow us to read a given ciphertext C which is assumed to be generated by putting an unknown plaintext P through an unknown monoalphabetic cipher.

We will denote the ciphertext alphabet as Σ_c and the plaintext alphabet as Σ_p. Given any string T, we will denote $n(T)$ as the length of T. Furthermore, we assume that the plaintext P is drawn from some string distribution q. We do not assume q to be a trigram distribution, but we do require it to be a distribution from which trigrams can be calculated (e.g, a 5-gram corpus will in general have probabilities that cannot be predicted using the associated trigrams, but the associated trigram corpus can be recovered from the 5-grams).

It is important to realize in the algorithm description and analysis that q may also not be known exactly, but we only assume that it exists, and that we can approximate it with a known trigram distribution p. In Corlett and Penn (2010), for example, p is the trigram distribution found using the Penn treebank. It is assumed that this is a good approximation for the distribution q, which in Corlett and Penn (2010) is the text in Wikipedia from which ciphers are drawn. As is common when dealing with probability distributions over natural languages, we assume that both p and q are stationary and ergodic, and we furthermore assume that p is smooth enough that any trigram that can be found in any string generated by q occurs in p (i.e., we assume that the cross entropy $H(p, q)$ is finite).

The algorithm works in a model in which, for any run of the algorithm, the plaintext string P is drawn according to the distribution q. We do not directly observe P, but instead its encoding using the cipher key, which we will call π_T. We observe the ciphertext $C = \pi_T^{-1}(P)$. We note that π_T is unknown, but that it does not change as new ciphertexts are drawn.

Now, the way that the algorithm in Corlett and Penn (2010) works is by searching over the possible keys to the cipher to find the one that maximizes the probability of the plaintext according to the distribution p. It does so as follows.

In addition to the possible keys to the cipher,

weakened cipher keys called *partial solutions* are added to the search space. A partial solution of size k (denoted as π_k) is a section of a possible full cipher key which is only defined on k character types in the cipher. We consider the character types to be fixed according to some preset order, and so the k fixed letters in π_k do not change between different partial solutions of size k.

Given a partial solution π_k, a string $\pi_k^{n(C)}(C)$ is defined whose probability we use as an upper bound for the probability of the plaintext whenever the true solution to the cipher contains π_k as a subset. The string $\pi_k^{n(C)}(C)$ is the most likely string that we can find that is consistent with C on the letters fixed by π_k. That is, we define the set Π_k so that $S \in \Pi_k$ iff whenever s_i and c_i are the characters at index i in S and C, then $s_i = \pi_k(c_i)$ if c_i is fixed in π_k. Note that if c_k is not fixed in π_k, we let s_i take any value. We extend the partial character function to the full string function $\pi_k^{n(C)}$ on $\Sigma_c^{n(C)}$ so that $\pi_k^{n(C)}(C) = argmax_{(S \in \Pi_k)} prob_p(S)$.

In Corlett and Penn (2010), the value $\pi_k^{n(C)}(C)$ is efficiently computed by running it through the Viterbi algorithm. That is, given C, p and π_k, a run of the Viterbi algorithm is set up in which the letter transition probabilities are those that are given in p. In order to describe the emission probabilities, suppose that we partition the ciphertext alphabet Σ_c into two sets Σ_1 and Σ_2, where Σ_1 is the set of ciphertext letters fixed by π_k. For any plaintext letter $y \in \Sigma_p$, if there is a ciphertext letter $x \in \Sigma_1$ such that $y \rightarrow x$ is a rule in π_k, then the emission probability that y will be seen as x is set to 1, and the probability that y will be seen as any other letter is set to 0. On the other hand, if there is no rule $y \rightarrow x$ in π_k for any ciphertext letter x, then the emission probability associated with y is uniform over the letters $x \in \Sigma_2$ and 0 for the letters $x \in \Sigma_1$.

The search algorithm described in Corlett and Penn (2010) uses the probability of the string $\pi_k^{n(C)}(C)$, or more precisely, the log probability $-logprob_p(\pi_k^{n(C)}(C))$, as an A^* heuristic over the partial solutions π_k. In this search, an edge is added from a size k partial solution π_k to a size $k+1$ partial solution π_{k+1} if π_k agrees with π_{k+1} wherever it is defined. The score of a node π_k is the log probability of its associated string: $-logprob_p(\pi_k^{n(C)}(C))$. We can see that if π_k has an edge leading to π_{k+1}, then $\Pi_{k+1} \subset \Pi_k$, so that $-logprob_p(\pi_{k+1}^{n(C)}(C)) \geq -logprob_p(\pi_k^{n(C)}(C))$. Thus, the heuristic is nondecreasing. Moreover, by applying the same statement inductively we can see that any full solution to the cipher that has π_k as a subset must have a score at least as great as that of π_k. This means that the score never overestimates the cost of completing a solution, and therefore that the heuristic is admissible.

4 Analysis

The bound that we will prove is that for any $k > 0$ and for any $\delta, \varepsilon > 0$, there exists an $n \in \mathbb{N}$ such that if the length $n(C)$ of the cipher C is at least n, then with probability at least $1 - \delta$, the search for the key to the cipher C requires no more than $2^{n \cdot (H(p,q)+\varepsilon)}$ expansions of any partial solution of size k to complete. Applying the same bound over every size k of partial solution will then give us that for any $\delta, \varepsilon > 0$, there exists a $n_0 > 0$ such that if the length $n(C)$ of the cipher C is at least n, then with probability at least $1 - \delta$, the search for the key to the cipher C requires no more than $2^{n(H(p,q)+\varepsilon)}$ expansions of any partial solution of size greater than 0 to complete (note that there is only one partial solution of size 0).

Let π^* be the solution that is found by the search. This solution has the property that it is the full solution that induces the most probable plaintext from the cipher, and so it produces a plaintext that is at least as likely as that of the true solution P. Thus, we have that $-logprob_p(\pi^{*n(C)}(C)) \leq -logprob_p(\pi_T^{n(C)}(C)) = -logprob_p(P)$.

We find our bound by making use of the fact that an A^* search never expands a node whose score is greater than that of the goal node π^*. Thus, a partial solution π_k is expanded only if

$$-logprob_p(\pi_k^{n(C)}(C)) \leq -logprob_p(\pi^{*n(C)}(C)).$$

Since

$$-logprob_p(\pi^{*n(C)}(C)) \leq -logprob_p(P),$$

we have that π_k is expanded only if

$$-logprob_p(\pi_k^{n(C)}(C)) \leq -logprob_p(P).$$

So we would like to count the number of solutions satisfying this inequality.

We would first like to approximate the value of $-logprob_p(P)$, then. But, since P is drawn from an ergodic stationary distribution q, this value will approach the cross entropy $H(p,q)$ with high probability: for any $\delta_1, \varepsilon_1 > 0$, there exists an $n_1 > 0$ such that if $n(C) = n(P) > N_1$, then

$$|-logprob_p(P)/n(C) - H(p,q)| < \varepsilon_1$$

with probability at least $1 - \delta_1$. In this case, we have that $-logprob_p(P) < n(C)(H(p,q) + \varepsilon_1)$.

Now, if k is fixed, and if π_k and π'_k are two different size k partial solutions, then π_k and π'_k must disagree on at least one letter assignment. Thus, the sets Π_k and Π'_k must be disjoint. But then we also have that $\pi_k^{n(C)}(C) \neq \pi_k^{n(C)\prime}(C)$. Therefore, if we can find an upper bound for the size of the set

$$\{S \in \Sigma_p^{n(C)} | S = \pi_k^{n(C)}(C) \text{ for some } \pi_k\},$$

we will have an upper bound on the number of times the search will expand any partial solution of size k. We note that under the previous assumptions, and with probability at least $1 - \delta_1$, none of these strings can have a log probability larger than $n(C)(H(p,q) + \varepsilon_1)$.

For any plaintext string C drawn from q, we let ${}_aP_b$ be the substring of P between the indices a and b. Similarly, we let ${}_aS_b$ be the substring of $S = \pi_k^{n(C)}(C)$ between the indices a and b.

We now turn to the proof of our bound: Let $\delta, \varepsilon > 0$ be given. We give the following three bounds on n:

(a) As stated above, we can choose n_1 so that for any string P drawn from q with length at least n_1,

$$|-logprob_p(P)/n(P) - H(p,q)| < \varepsilon_1/2$$

with probability at least $1 - \delta/3$.

(b) We have noted that if k is fixed then any two size k partial solutions must disagree on at least one of the letters that they fix. So if we have a substring ${}_aP_b$ of P with an instance of every letter type fixed by the partial solutions of size k, then the substrings ${}_aS_b$ of S must be distinct for every $S \in \{S \in \Sigma_p^{n(C)} | S = \pi_k^{n(C)}(C)$ for some $\pi_k\}$. Since q is ergodic, we can find an n_2 such that for any string P drawn from q with length at least n_2, every letter fixed in π_k can be found in some length n_2 substring P_2 of P, with probability at least $1 - \delta/3$.

(c) By the Lemma below, there exists an $n' > 0$ such that for all partial solutions π_k, there exists a trigram distribution r_k on the alphabet Σ_p such that if $S = \pi_k^{n(C)}(C)$ and $b - a = n > n'$, then

$$\left|\frac{-logprob({}_aS_b)}{n} - H(p, r_k)\right| < \varepsilon/4$$

with a probability of at least $1 - \delta/3$.

Let $n = \max(n_1, n_2, n')$. Then, the probability of any single one of the properties in (a), (b) or (c) failing in a string of length at least n is at most $\delta/3$, and so the probability of any of them failing is at most δ. Thus, with a probability of at least $1-\delta$, all three of the properties hold for any string P drawn from q with length at least n. Let P be drawn from q, and suppose $n(P) > n$. Let ${}_aP_b$ be a length n substring of P containing a token of every letter type fixed by the size k partial solutions.

Suppose that π_k is a partial solution such that $-logprob_p(\pi_k^{n(C)}(C)) \leq n(P)(H(p,q) + \varepsilon/2)$. Then, letting $S = \pi_k^{n(C)}(C)$, we have that if

$$\left|\frac{-logprob(S)}{n(P)} - H(p, r_k)\right| < \varepsilon/4$$

and

$$\left|\frac{-logprob({}_aS_b)}{n} - H(p, r_k)\right| < \varepsilon/4$$

it follows that

$$\begin{aligned}
&\left|\frac{-logprob(S)}{n(P)} + \frac{logprob({}_aS_b)}{n}\right| \\
&\leq \left|\frac{-logprob(S)}{n(P)} - H(p, r_k)\right| \\
&+ \left|-H(p, r_k) - \frac{logprob({}_aS_b)}{n}\right| \\
&\leq \varepsilon/4 + \varepsilon/4 = \varepsilon/2
\end{aligned}$$

But then,

$$\begin{aligned}
-\frac{logprob({}_aS_b)}{n} &< \frac{-logprob(S)}{n(P)} + \varepsilon/2 \\
&\leq \frac{n(P)(H(p,q) + \varepsilon/2)}{n(P)} + \varepsilon/2 \\
&= H(p,q) + \varepsilon.
\end{aligned}$$

So, for our bound we will simply need to find the number of substrings ${}_aS_b$ such that

$$-\log prob_p({}_aS_b) < n(H(p,q)+\varepsilon).$$

Letting $I_H({}_aS_b) = 1$ if $-logprob_p({}_aS_b) < n(H(p,q)+\varepsilon)$ and 0 otherwise, the number of strings we need becomes

$$\begin{aligned}\sum_{{}_aS_b\in\Sigma_p^{n(C)}} I_H({}_aS_b) &= 2^{n\cdot(H(p,q)+\varepsilon)}\sum_{{}_aS_b\in\Sigma_p^{n(C)}} I_H({}_aS_b)2^{-n\cdot(H(p,q)+\varepsilon)}\\ &< 2^{n\cdot(H(p,q)+\varepsilon)}\sum_{{}_aS_b\in\Sigma_p^{n(C)}} I_H({}_aS_b)prob_p({}_aS_b)\\ &\text{(since } -\log prob_p({}_aS_b) < n(H(p,q)+\varepsilon)\\ &\text{implies } prob_p({}_aS_b) > 2^{-n\cdot(H(p,q)+\varepsilon)})\\ &\leq 2^{n\cdot(H(p,q)+\varepsilon)}\sum_{{}_aS_b\in\Sigma_p^{n(C)}} prob_p({}_aS_b)\\ &= 2^{n\cdot(H(p,q)+\varepsilon)}\end{aligned}$$

Thus, we have a bound of $2^{n\cdot(H(p,q)+\varepsilon)}$ on the number of substrings of length n satisfying $-\log prob_p({}_aS_b) < n(H(p,q)+\varepsilon)$. Since we know that with probability at least $1-\delta$, these are the only strings that need be considered, we have proven our bound. □

4.1 Lemma:

We now show that for any fixed $k > 0$ and $\delta', \varepsilon' > 0$, there exists some $n' > 0$ such that for all partial solutions π_k, there exists a trigram distribution r_k on the alphabet Σ_p such that if $S = \pi_k^{n(C)}(C)$ and $b - a = n > n', |\frac{-logprob({}_aS_b)}{n} - H(p, r_k)| < \varepsilon'$ with a probability of at least $1 - \delta'$.

Proof of Lemma: Given any partial solution π_k, it will be useful in this section to consider the strings $S = \pi_k^{n(C)}(C)$ as functions of the plaintext P rather than the ciphertext C. Since $C = \pi_T^{-1}(P)$, then, we will compose $\pi_k^{n(C)}$ and π_T^{-1} to get $\pi_k^{n(C)\prime}(P) = \pi_k^{n(C)}(\pi_T^{-1}(P))$. Now, since π_T is derived from a character bijection between Σ_c and Σ_p, and since $\pi_k^{n(C)}$ fixes the k character types in Σ_c that are defined in π_k, we have that $\pi_k^{n(C)\prime}$ fixes k character types in Σ_p. Let Σ_{P_1} be the set of k character types in Σ_p that are fixed by $\pi_k^{n(C)\prime}$, and let $\Sigma_{P_2} = \Sigma_p \setminus \Sigma_{P_1}$. We note that Σ_{P_1} and Σ_{P_2} do not depend on which π_k we use, but only on k.

Now, any string P which is drawn from q can be decomposed into overlapping substrings by splitting it whenever it has see two adjacent characters from Σ_{P_1}. When we see a bigram in P of this form, say, y_1y_2, we split P so that both the end of the initial string and the beginning of the new string are y_1y_2. Note that when we have more than two adjacent characters from Σ_{P_1} we will split the string more than once, so that we get a series of three-character substrings of P in our decomposition. As a matter of bookkeeping we will consider the initial segment to begin with two start characters **s** with indices corresponding to 0 and -1 in P. As an example, consider the string

P = *friends, romans, countrymen, lend me your ears*

Where $\Sigma_{P_1} = \{\text{' '}, \text{','}, \text{'}a\text{'}, \text{'}y\text{'}\}$. In this case, we would decompose P into the strings *'ssfriends, ', ', romans, ', ', countrymen, ', ', lend me ', 'e y', ' your e'* and *' ears'*.

Let M be the set of all substrings that can be generated in this way by decomposing strings P which are drawn from q. Since the end of any string $m \in M$ contains two adjacent characters in Σ_{P_1} and since the presence of two adjacent characters in Σ_{P_1} signals a position at which a string will be decomposed into segments, we have that the set M is prefix-free. Every string $m \in M$ is a string in Σ_p, and so they will have probabilities $prob_q(m)$ in q. It should be noted that for any $m \in M$ the probability $prob_q(m)$ may be different from the trigram probabilities predicted by q, but will instead be the overall probability in q of seeing the string m.

For any pair T, P of strings, let $\#(T,P)$ be the number of times T occurs in P. Since we assume that the strings drawn from q converge to the distribution q, we have that for any $\delta_3, \varepsilon_3 > 0$ and any $n_4 > 0$, there exists an $n_3 > 0$ such that for any substring P_3 of P of length at least n_3, where P is drawn from q, and for any $m \in M$ of length at most n_4, the number $|\#(m,P)/len(P_3) - prob_q(m)| < \varepsilon_3$ with probability greater than $1 - \delta_3$.

Now suppose that for some P drawn from q we have a substring ${}_aP_b$ of P such that ${}_aP_b = m, m \in M$. If $S = \pi_k^{n(C)\prime}(P)$, consider the substring ${}_aS_b$ of S. Recall that the string function $\pi_k^{n(C)\prime}$ can map the characters in P to S in one of two ways: if a character $x_i \in \Sigma_{P_1}$ is found at index i in P, then the corresponding character in S

is $\pi_k(x_i)$. Otherwise, x_i is mapped to whichever character y_i in Σ_P maximizes the probability in p of S given $\pi_k^{n(C)'}(x_{i-2})\pi_k^{n(C)'}(x_{i-1})y_i$. Since the values of $\pi_k^{n(C)'}(x_{i-2})$, $\pi_k^{n(C)'}(x_{i-1})$ and y_i are interdependent, and since $\pi_k^{n(C)'}(x_{i-2})$ is dependent on its previous two neighbors, the value that y_i takes may be dependent on the values taken by $\pi_k^{n(C)'}(x_j)$ for indices j quite far from i. However, we see that no dependencies can cross over a substring in P containing two adjacent characters in Σ_{P_1}, since these characters are not transformed by $\pi_k^{n(C)'}$ in a way that depends on their neighbors. Thus, if ${}_aP_b = m \in M$, the endpoints of ${}_aP_b$ are made up of two adjacent characters in Σ_{P_1}, and so the substring ${}_aS_b$ of S depends only on the substring ${}_aP_b$ of P. Specifically, we see that ${}_aS_b = \pi_k^{n(C)'}({}_aP_b)$.

Since we can decompose any P into overlapping substrings $m_1, m_2, \ldots, m_t$ in M, then, we can carry over this decomposition into S to break S into $\pi_k^{n(C)'}(m_1), \pi_k^{n(C)'}(m_2), \ldots, \pi_k^{n(C)'}(m_t)$. Note that the score generated by S in the A^* search algorithm is the sum $\sum_{1\le i\le} logprob_p(y_{i-2}y_{i-1}y_i)$, where y_i is the ith character in S. Also note that every three-character sequence $y_{i-2}y_{i-1}y_i$ occurs exactly once in the decomposition $\pi_k^{n(C)'}(m_1), \pi_k^{n(C)'}(m_2), \ldots, \pi_k^{n(C)'}(m_t)$. Since for any m the number of occurrences of $\pi_k^{n(C)'}(m)$ in S under this decomposition will be equal to the number of occurrences of m in P, we have that

$$\begin{aligned}
-logprob_p(S) &= \sum_{1\le i\le n(P)} logprob_p(y_{i-2}y_{i-1}y_i) \\
&= \sum_{m\in M} \#(m,P)\cdot(-logprob_p(\pi_k^{n(C)'}(m))).
\end{aligned}$$

Having finished these definitions, we can now define the distribution r_k. In principle, this distribution should be the limit of the frequency of trigram counts of the strings $S = \pi_k^{n(C)'}(P)$, where $n(P)$ approaches infinity. Given a string $S = \pi_k^{n(C)'}(P)$, where P is drawn from q, and given any trigram $y_1y_2y_3$ of characters in Σ_p, this frequency count is $\frac{\#(y_1y_2y_3,S)}{n(P)}$. Breaking S into its component substrings $\pi_k^{n(C)'}(m_1), \pi_k^{n(C)'}(m_2), \ldots, \pi_k^{n(C)'}(m_t)$, as we have done above, we see that any instance of the trigram $y_1y_2y_3$ in S occurs in exactly one of the substrings $\pi_k^{n(C)'}(m_i), 1 \le i \le t$. Grouping together similar m_is, we find

$$\begin{aligned}
\frac{\#(y_1y_2y_3,S)}{n(P)} &= \frac{\sum_{i=1}^{t}\#(y_1y_2y_3,\pi_k^{n(C)'}(m_i))}{n(P)} \\
&= \frac{\sum_{m\in M}\#(y_1y_2y_3,\pi_k^{n(C)'}(m))\cdot\#(m,P)}{n(P)}
\end{aligned}$$

As $n(P)$ approaches infinity, we find that $\frac{\#(m,P)}{n(P)}$ approaches $prob_q(m)$, and so we can write

$$prob_{r_k}(y_1y_2y_3) = \sum_{m\in M}\#(y_1y_2y_3,\pi_k^{n(C)'}(m))prob_q(m).$$

Since $0 \le \sum_{m\in M}\#(y_1y_2y_3,\pi_k^{n(C)'}(m))prob_q(m)$ when P is sampled from q we have that

$$\begin{aligned}
&\sum_{y_1y_2y_3} prob_{r_k}(y_1y_2y_3) \\
&= \sum_{y_1y_2y_3}\sum_{m\in M}\#(y_1y_2y_3,\pi_k^{n(C)'}(m))prob_q(m) \\
&= \lim_{n(P)\to\infty}\sum_{y_1y_2y_3}\sum_{m\in M}\#(y_1y_2y_3,\pi_k^{n(C)'}(m))\frac{\#(m,P)}{n(P)} \\
&= \lim_{n(P)\to\infty}\sum_{m\in M}\sum_{y_1y_2y_3}\#(y_1y_2y_3,\pi_k^{n(C)'}(m))\frac{\#(m,P)}{n(P)} \\
&= \lim_{n(P)\to\infty}\sum_{m\in M}\frac{(n(\pi_k^{n(C)'}(m))-2)\#(m,P)}{n(P)} \\
&= \lim_{n(P)\to\infty}\sum_{m\in M}\frac{(n(m)-2)\#(m,P)}{n(P)} \\
&= \lim_{n(P)\to\infty}\frac{n(P)}{n(P)} = 1,
\end{aligned}$$

so we have that $prob_{r_k}$ is a valid probability distribution. In the above calculation we can rearrange the terms, so convergence implies absolute convergence. The sum $\sum_{y_1y_2y_3}\#(y_1y_2y_3,\pi_k^{n(C)'}(m))$ gives $(n(\pi_k^{n(C)'}(m))-2)$ because there is one trigram for every character in $\pi_k^{n(C)'}(m)$, less two to compensate for the endpoints. However, since the different m overlap by two in a decomposition from P, the sum $(n(m)-2)\#(m,P)$ just gives back the length $n(P)$, allowing for the fact that the initial m has two extra start characters.

Having defined r_k, we can now find the value of $H(p, r_k)$. By definition, this term will be

$$\sum_{y_1y_2y_3} -logprob_p(y_1y_2y_3)prob_{r_k}(y_1y_2y_3)$$
$$= \sum_{y_1y_2y_3} -logprob_p(y_1y_2y_3)\sum_{m\in M}\#(y_1y_2y_3, \pi_k^{n(C)\prime}(m))prob_q(m)$$
$$= \sum_{m\in M}\sum_{y_1y_2y_3} -logprob_p(y_1y_2y_3)\#(y_1y_2y_3, \pi_k^{n(C)\prime}(m))prob_q(m)$$
$$= \sum_{m\in M} -logprob_p(m)prob_q(m).$$

Now, we can finish the proof of the Lemma. Holding k fixed, let $\delta', \varepsilon' > 0$ be given. Since we have assumed that p does not assign a zero probability to any trigram generated by q, we can find a trigram $x_1x_2x_3$ generated by q whose probability in p is minimal. Let $X = -logprob_p(x_1x_2x_3)$, and note that $prob_p(x_1x_2x_3) > 0$ implies $X < \infty$. Since we know by the argument above that when P is sampled from q, $\lim_{n(P)\to\infty}(\sum_{m\in M} \frac{(n\pi_k^{n(C)\prime}(m)-2)\cdot\#(m,P)}{n(P)}) = 1$, we have that

$$\sum_{m\in M} (n\pi_k^{n(C)\prime}(m) - 2)prob_q(m) = 1.$$

Thus, we can choose n_4 so that

$$\sum_{m\in M, n(m)\le n_4} (n\pi_k^{n(C)\prime}(m) - 2)prob_q(m) > 1 - \varepsilon'/4X.$$

Let $Y = |\{m \in M, n(m) \le n_4\}|$, and choose n' such that if P is sampled from q and ${}_aP_b$ is a substring of P with length greater than n', then with probability at least $1 - \delta'$, for every $m \in M$ we will have that

$$\left|\frac{\#(m, {}_aP_b)}{n({}_aP_b)} - prob_q(m)\right| < \varepsilon'/4XY(n_4 - 2).$$

Let π_k be any partial solution of length k, and let r_k be the trigram probability distribution described above. Then let P be sampled from q, and let $S = \pi_k^{n(C)}(C) = \pi_k^{n(C)\prime}(P)$, and let a, b be indices of S such that $b - a = n > n'$. Finally, we will partition the set M as follows: we let M' be the set $\{m \in M | n(n) \le n_4\}$ and M'' be the set $\{m \in M | n(m) > n_4\}$. Thus, we have that

$$\left|\frac{-logprob({}_aS_b)}{n} - H(p, r_k)\right| = \left|\frac{\sum_{m\in M}\#(m, {}_aP_b)(-logprob_p(\pi_k^{n(C)\prime}(m))}{n} - \sum_{m\in M} prob_q(m)\cdot(-logprob_p(\pi_k^{n(C)\prime}(m))\right|.$$

Grouping the terms of these sums into the index sets M' and M'', we find that this value is at most

$$\left|\sum_{m\in M'}\left(\frac{\#(m, {}_aP_b)}{n} - prob_q(m)\right)(-logprob_p(\pi_k^{n(C)\prime}(m))\right| + \left|\sum_{m\in M''}\left(\frac{\#(m, {}_aP_b)}{n} - prob_q(m)\right)(-logprob_p(\pi_k^{n(C)\prime}(m))\right|$$

Furthermore, we can break up the sum over the index M'' to bound this value by

$$\left|\sum_{m\in M'}\left(\frac{\#(m, {}_aP_b)}{n} - prob_q(m)\right)(-logprob_p(\pi_k^{n(C)\prime}(m))\right| + \left|\sum_{m\in M''}\frac{\#(m, {}_aP_b)}{n}(-logprob_p(\pi_k^{n(C)\prime}(m))\right| + \left|\sum_{m\in M''} prob_q(m)(-logprob_p(\pi_k^{n(C)\prime}(m))\right|$$

Now, for any $m \in M$, we have that the score $-logprob_p(\pi_k^{n(C)\prime}(m)$ equals $\sum_{1\le i\le n(m)-2} -logprob_p(y_iy_{i+1}y_{i+2})$, where y_i is the character at the index i in $\pi_k^{n(C)\prime}(m)$. Taking the maximum possible values for $-logprob_p(y_iy_{i+1}y_{i+2})$, we find that this sum is at most $(n(m) - 2)X$. Applying this bound to the previous formula, we find that it is at most

$$\left|\sum_{m\in M'}\left(\frac{\#(m, {}_aP_b)}{n} - prob_q(m)\right)(n(m) - 2)X\right| + \left|\sum_{m\in M''}\frac{\#(m, {}_aP_b)}{n}(n(m) - 2)X\right| + \left|\sum_{m\in M''} prob_q(m)\cdot(n(m) - 2)X\right|.$$

We can bound each of these three terms separately. Looking at the first sum in this series, we find that with probability at least $1 - \delta'$,

$$\left|\sum_{m\in M'}\left(\frac{\#(m,{}_aP_b)}{n} - prob_q(m)\right)(n(m)-2)X\right| \quad (*)$$
$$\leq \sum_{m\in M'}\left|\frac{\#(m,{}_aP_b)}{n} - prob_q(m)\right|(n(m)-2)X$$
$$\leq \sum_{m\in M'}\left|\frac{\varepsilon'}{4(n_4-2)XY}\right|\cdot(n(m)-2)X$$
$$\leq \sum_{m\in M'}\left|\frac{\varepsilon'}{4Y}\right|$$
$$= \frac{\varepsilon'}{4Y}\sum_{m\in M'}1 = \frac{\varepsilon'}{4Y}Y = \varepsilon/4.$$

In order to bound the second sum, we make use of the fact that $\sum_{m\in M}\#(m,{}_aP_b)(n(m)-2) = n({}_aP_b) = n$ to find that once again, with probability greater than $1-\delta'$,

$$\left|\sum_{m\in M''}\frac{\#(m,{}_aP_b)}{n}(n(m)-2)X\right|$$
$$\leq \sum_{m\in M''}\left|\frac{\#(m,{}_aP_b)}{n}(n(m)-2)X\right|.$$

Since $M'' = M - M'$, this value is

$$\sum_{m\in M}\left|\frac{\#(m,{}_aP_b)}{n}(n(m)-2)X\right|$$
$$-\sum_{m\in M'}\left|\frac{\#(m,{}_aP_b)}{n}(n(m)-2)X\right|$$
$$=X-\sum_{m\in M'}\left|\frac{\#(m,{}_aP_b)}{n}(n(m)-2)X\right|.$$

This value can further be split into

$$=X-\sum_{m\in M'}\left|\left(\frac{\#(m,{}_aP_b)}{n}+(1-1)prob_q(m)\right)(n(m)-2)X\right|$$
$$\leq X-\left(\sum_{m\in M'}|prob_q(m)(n(m)-2)X|\right.$$
$$\left.-\sum_{m\in M'}\left|\frac{\#(m,{}_aP_b)}{n}-prob_q(m)\right|(n(m)-2)X\right)$$

Using our value for the sum in (*), we find that this is

$$=X-\sum_{m\in M'}|prob_q(m)(n(m)-2)X|$$
$$+\sum_{m\in M'}\left|\frac{\#(m,{}_aP_b)}{n}-prob_q(m)\right|(n(m)-2)X$$
$$\leq X-\sum_{m\in M'}|prob_q(m)(n(m)-2)X|+\frac{\varepsilon'}{4},$$

Using our definition of n_4, we can further bound this value by

$$=X\left(1-\sum_{m\in M'}prob_q(m)(n(m)-2)\right)+\frac{\varepsilon'}{4}$$
$$<X\left(1-\left(1-\frac{\varepsilon'}{4X}\right)\right)+\frac{\varepsilon'}{4}$$
$$=X\frac{\varepsilon'}{4X}+\frac{\varepsilon'}{4}=\frac{\varepsilon'}{2}.$$

Finally, we once again make use of the definition of n_4 to find that the last sum is

$$\left|\sum_{m\in M''}prob_q(m)\cdot(n(m)-2)X\right|$$
$$=\sum_{m\in M''}prob_q(m)\cdot(n(m)-2)X$$
$$=X\sum_{m\in M''}prob_q(m)\cdot(n(m)-2)$$
$$<X\frac{\varepsilon'}{4X}$$
$$=\frac{\varepsilon'}{4}.$$

Adding these three sums together, we get

$$\frac{\varepsilon'}{4}+\frac{\varepsilon'}{2}+\frac{\varepsilon'}{4}=\varepsilon'.$$

Thus, $\left|\frac{-logprob({}_aS_b)}{n}-H(p,r_k)\right|<\varepsilon'$ with probability greater than $1-\delta'$, as required. □

5 Conclusion

In this paper, we discussed a discrepancy between the theoretical and practical running times of certain algorithms that are sensitive to the entropies of their input, or the entropies of the distributions from which their inputs are sampled. We then used the algorithm from Corlett and Penn (2010) as a subject to allow us to investigate ways to talk about average-case complexity in light of this discrepancy. Our analysis was sufficient to give us a bound on the search complexity of this algorithm which is exponential in the cross-entropy between the training distribution and the input distribution. Our method in effect yields a probabilistic bound on the depth of the search heuristic used. This leads to an exponentially smaller search space for the overall problem.

We must note, however, that our analysis does not fully reconcile the discrepancy between the theoretical and practical running time for this algorithm. In particular, our bound still does not explain why the number of search nodes expanded by this algorithm tends to converge on one per partial solution size as the length of the string grows very large. As such, we are interested in further studies as to how to explain the running time of this algorithm. It is our opinion that this can be done by refining our description of the sets Π_k to exclude strings which cannot be considered by the algorithm. Not only would this allow us to reduce the overall number of strings we would have to count when determining the bound, but we would also have to consider fewer strings when determining the value of n'. Both changes would reduce the overall complexity of our bound.

This general strategy may have the potential to illuminate the practical time complexities of approximate search algorithms as well.

References

David Arthur, Bodo Manthey, and Heiko Röglin. k-means has polynomial smoothed complexity. In *The 50th Annual Symposium on Foundations of Computer Science*. IEEE Computer Society Technical Committee on Mathematical Foundations of Computing, 2009. URL `http://arxiv.org/abs/0904.1113`.

Eric Corlett and Gerald Penn. An exact A* method for deciphering letter-substitution ciphers. In *Proceedings of the 48th Annual Meeting of the Association for Computational Linguistics*, pages 1040–1047, 2010.

Richard E Korf and Michael Reid. Complexity analysis of admissible heuristic search. In *Proceedings of the Fifteenth National Conference on Artificial Intelligence*, 1998.

Richard E Korf, Michael Reid, and Stefan Edelkamp. Time complexity of iterative-deepening-A*. *Artificial Intelligence*, 129(1–2): 199–218, 2001.

Judea Pearl. *Heuristics: Intelligent Search Strategies for Computer Problem Solving*. Addison-Wesley, 1984.

Daniel A Spielman and Shang-Hua Teng. Smoothed analysis of algorithms: Why the simplex algorithm usually takes polynomial time. *Journal of the ACM*, 51(3):385–463, 2004.

Investigating Connectivity and Consistency Criteria for Phrase Pair Extraction in Statistical Machine Translation

Spyros Martzoukos, Christophe Costa Florêncio and Christof Monz
Intelligent Systems Lab Amsterdam, University of Amsterdam
Science Park 904, 1098 XH Amsterdam, The Netherlands
{S.Martzoukos, C.Monz}@uva.nl, chriscostafl@gmail.com

Abstract

The consistency method has been established as the standard strategy for extracting high quality translation rules in statistical machine translation (SMT). However, no attention has been drawn to why this method is successful, other than empirical evidence. Using concepts from graph theory, we identify the relation between consistency and components of graphs that represent word-aligned sentence pairs. It can be shown that phrase pairs of interest to SMT form a sigma-algebra generated by components of such graphs. This construction is generalized by allowing segmented sentence pairs, which in turn gives rise to a phrase-based generative model. A by-product of this model is a derivation of probability mass functions for random partitions. These are realized as cases of constrained, biased sampling without replacement and we provide an exact formula for the probability of a segmentation of a sentence.

1 Introduction

A parallel corpus, i.e., a collection of sentences in a source and a target language, which are translations of each other, is a core ingredient of every SMT system. It serves the purpose of training data, i.e., data from which translation rules are extracted. In its most basic form, SMT does not require the parallel corpus to be annotated with linguistic information, and human supervision is thus restricted to the construction of the parallel corpus.

The extraction of translation rules is done by appropriately collecting statistics from the training data. The pioneering work of (Brown et al., 1993) identified the minimum assumptions that should be made in order to extract translation rules and developed the relevant models that made such extractions possible.

These models, known as IBM models, are based on standard machine learning techniques. Their output is a matrix of word alignments for each sentence pair in the training data. These word alignments provide the input for later approaches that construct phrase-level translation rules which may (Wu, 1997; Yamada and Knight, 2001) or may not (Och et al., 1999; Marcu and Wong, 2002) rely on linguistic information.

The method developed in (Och et al., 1999), known as the *consistency* method, is a simple yet effective method that has become the standard way of extracting (source, target)-pairs of phrases as translation rules. The development of consistency has been done entirely on empirical evidence and it has thus been termed a heuristic.

In this work we show that the method of (Och et al., 1999) actually encodes a particular type of structural information induced by the word alignment matrices. Moreover, we show that the way in which statistics are extracted from the associated phrase pairs is insufficient to describe the underlying structure.

Based on these findings we suggest a phrase-level model in the spirit of the IBM models. A key aspect of the model is that it identifies the most likely partitions, rather than alignment maps, associated with appropriately chosen segments of the training data. For that reason, we provide a *general* construction of probability mass functions for partitions and, in particular, an exact formula for the probability of a segmentation of a sentence.

2 Definition of Consistency

In this section we provide the definition of consistency, which was introduced in (Och et al., 1999), refined in (Koehn et al., 2003), and we follow (Koehn, 2009) in our description. We start with some preliminary definitions.

Let $S = s_1...s_{|S|}$ be a source sentence, i.e., a string that consists of consecutive source words; each word s_i is drawn from a source language vocabulary and i indicates the position of the word in S. The operation of string *extraction* from the words of S is defined as the construction of the string $s = s_{i_1}...s_{i_n}$ from the words of S, with $1 \leq i_1 < ... < i_n \leq |S|$. If $i_1, ..., i_n$ are consecutive, which implies that s is a substring of S, then s is called a source phrase and we write $s \subseteq S$. As a shorthand we also write $s_{i_1}^{i_n}$ for the phrase $s_{i_1}...s_{i_n}$. Similar definitions apply to the target side and we denote by T, t_j and t a target sentence, word and phrase respectively.

Let $(S = s_1 s_2...s_{|S|}, T = t_1 t_2...t_{|T|})$ be a sentence pair and let A denote the $|S| \times |T|$ matrix that encodes the existence/absence of word alignments in (S, T) as

$$A(i,j) = \begin{cases} 1, & \text{if } s_i \text{ and } t_j \text{ are aligned} \\ 0, & \text{otherwise,} \end{cases} \quad (1)$$

for all $i = 1, ..., |S|$ and $j = 1, ..., |T|$. Unaligned words are allowed. A pair of strings $(s = s_{i_1}...s_{i_{|s|}}, t = t_{j_1}...t_{j_{|t|}})$ that is extracted from (S, T) is termed *consistent* with A, if the following conditions are satisfied:

1. $s \subseteq S$ and $t \subseteq T$.
2. $\forall k \in \{1, ..., |s|\}$ such that $A(i_k, j) = 1$, then $j \in \{j_1, ..., j_{|t|}\}$.
3. $\forall l \in \{1, ..., |t|\}$ such that $A(i, j_l) = 1$, then $i \in \{i_1, ..., i_{|s|}\}$.
4. $\exists k \in \{1, ..., |s|\}$ and $\exists l \in \{1, ..., |t|\}$ such that $A(i_k, j_l) = 1$.

Condition 1 guarantees that (s, t) is a phrase pair and not just a pair of strings. Condition 2 says that if a word in s is aligned to one or more words in T, then all such target words must appear in t. Condition 3 is the equivalent of Condition 2 for the target words. Condition 4 guarantees the existence of at least one word alignment in (s, t).

For a sentence pair (S, T), the set of all consistent pairs with an alignment matrix A is denoted by $P(S, T)$. Figure 1(a) shows an example of a sentence pair with an alignment matrix together with all its consistent pairs.

In SMT the extraction of each consistent pair (s, t) from (S, T) is followed by a statistic $f(s, t; S, T)$. Typically $f(s, t; S, T)$ counts the occurrences of (s, t) in (S, T). By considering all sentence pairs in the training data, the translation probability is constructed as

$$p(t|s) = \frac{\sum_{(S,T)} f(s, t; S, T)}{\sum_{(S,T)} \sum_{t'} f(s, t'; S, T)}, \quad (2)$$

and similarly for $p(s|t)$. Finally, the entries of the phrase table consist of all extracted phrase pairs, their corresponding translation probabilities and other models which we do not discuss here.

3 Consistency and Components

For a given sentence pair (S, T) and a fixed word alignment matrix A, our aim is to show the equivalence between consistency and connectivity properties of the graph formed by (S, T) and A. Moreover, we explain that the way in which measurements are performed is not compatible , in principle, with the underlying structure. We start with some basic definitions from graph theory (see for example (Harary, 1969)).

Let $G = (V, E)$ be a graph with vertex set V and edge set E. Throughout this work, vertices represent words and edges represent word alignments, but the latter will be further generalized in Section 4. A *subgraph* $H = (V', E')$ of G is a graph with $V' \subseteq V$, $E' \subseteq E$ and the property that for each edge in E', both its endpoints are in V'. A *path* in G is a sequence of edges which connect a sequence of distinct vertices. Two vertices $u, v \in V$ are called connected if G contains a path from u to v. G is said to be *connected* if every pair of vertices in G is connected.

A connected component, or simply *component*, of G is a maximal connected subgraph of G. G is called *bipartite* if V can be partitioned in sets V_S and V_T, such that every edge in E connects a vertex in V_S to one in V_T. The disjoint union of graphs, or simply *union*, is an operation on graphs defined as follows. For n graphs with disjoint vertex sets $V_1, ..., V_n$ (and hence disjoint edge sets), their union is the graph $(\cup_{i=1}^n V_i, \cup_{i=1}^n E_i)$.

Consider the graph G whose vertices are the words of the source and target sentences, and whose edges are induced by the non-zero entries

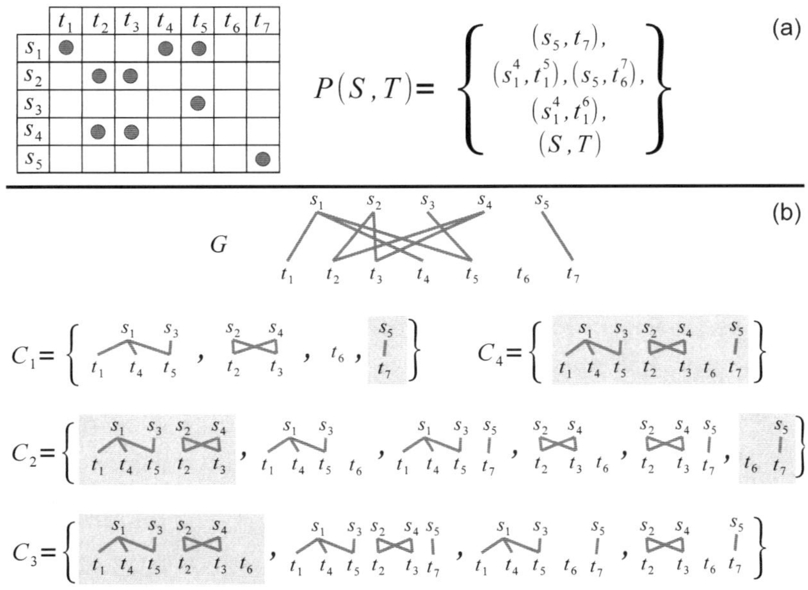

Figure 1: (a) Left: Sentence pair with an alignment matrix. Dots indicate existence of word alignments. Right: All consistent pairs. (b) The graph representation of the matrix in (a), and the sets generated by components of the graph. Dark shading indicates consistency.

of the matrix A. There are no edges between any two source-type vertices nor between any two target-type vertices. Moreover, the source and target language vocabularies are assumed to be disjoint and thus G is bipartite. The set of all components of G is defined as C_1 and let k denote its cardinality, i.e., $|C_1| = k$. From the members of C_1 we further construct sets $C_2, ..., C_k$ as follows: For each i, $2 \leq i \leq k$, any member of C_i is formed by the union of any i distinct members of C_1. In other words, any member of C_i is a graph with i components and each such component is a member of C_1. The cardinality of C_i is clearly $\binom{k}{i}$, for every i, $1 \leq i \leq k$.

Note that $C_k = \{G\}$, since G is the union of all members of C_1. Moreover, observe that $C_* = \cup_{i=1}^{k} C_i$ is the set of graphs that can be generated by all possible unions of G's components. In that sense

$$C = \{\emptyset\} \cup C_* \tag{3}$$

is the power set of G. Indeed we have $|C| = 1 + \sum_{i=1}^{k} \binom{k}{i} = 2^k$ as required.[1]

Figure 1(b) shows the graph G and the associated sets C_i of (S, T) and A in Figure 1(a). Note the bijective correspondence between consistent pairs and the phrase pairs that can be extracted from the vertices of the members of the sets C_i. This is a consequence of consistency Conditions 2 and 3, since they provide the sufficient conditions for component formation.

In general, if a pair of strings (s, t) satisfies the consistency Conditions 2 and 3, then it can be extracted from the vertices of a graph in C_i, for some i. Moreover, if Conditions 1 and 4 are also satisfied, i.e., if (s, t) is consistent, then we can write

$$P(S,T) = \bigcup_{i=1}^{k} \{\, (S_H, T_H) : \; H \in C_i, \quad S_H \subseteq S, \; T_H \subseteq T \,\}, \tag{4}$$

where S_H denotes the extracted string from the source-type vertices of H, and similarly for T_H. Having established this relationship, when referring to members of C, we henceforth mean either consistent pairs or *inconsistent* pairs. The latter are pairs (S_H, T_H) for some $H \in C$ such that at least either $S_H \not\subseteq S$ or $T_H \not\subseteq T$.

The construction above shows that phrase pairs of interest to SMT are part of a carefully constructed subclass of all possible string pairs that can be extracted from (S, T). The power set C of G gives rise to a small, possibly minimal, set

[1] Here we used the fact that for any set X with $|X| = n$, the set of all subsets of X, i.e., the power set of X, has cardinality $\sum_{i=0}^{n} \binom{n}{i} = 2^n$.

in which consistent and inconsistent pairs can be *measured*.[1] In other words, since C is (by construction) a sigma-algebra, the pair (C_1, C) is a measurable space. Furthermore, one can construct a measure space (C_1, C, f), with an appropriately chosen measure $f : C \to [0, \infty)$.

Is the occurrence-counting measure f of Section 2 a good choice? Fix an ordering for C_i, and let $C_{i,j}$ denote the jth member of C_i, for all i, $1 \leq i \leq k$. Furthermore, let $\delta(x, y) = 1$, if $x = y$ and 0, otherwise. We argue by contradiction that the occurrence-counting measure

$$f(H) = \sum_{\{H':\ H' \in C,\ H' \text{ is consistent}\}} \delta(H, H'), \quad (5)$$

fails to form a measure space. Suppose that more than one component of G is consistent, i.e., suppose that

$$1 < \sum_{j=1}^{k} f(C_{1,j}) \leq k. \quad (6)$$

By construction of C, it is guaranteed that

$$1 = f(G) = f(C_{k,1}) = f(\cup_{j=1}^{k} C_{1,j}). \quad (7)$$

The members of C_1 are pairwise disjoint, because each of them is a component of G. Thus, since f is assumed to be a measure, sigma-additivity should be satisfied, i.e., we must have

$$f(\cup_{j=1}^{k} C_{1,j}) = \sum_{j=1}^{k} f(C_{1,j}) > 1, \quad (8)$$

which is a contradiction.

In practice, the deficiency of using eq. 5 as a statistic could possibly be explained by the fact that the so-called lexical weights are used as smoothing.

4 Consistency, Components and Segmentations

In Section 3 the only relation that was assumed among source (target) words/vertices was the order of appearance in the source (target) sentence. As a result, the graph representation G of (S, T) and A was bipartite. There are several, linguistically motivated, ways in which a general graph can be obtained from the bipartite graph G. We explain that the minimal linguistic structure, namely sentence segmentations, can provide a generalization of the construction introduced in Section 3.

Let X be a finite set of consecutive integers. A consecutive partition of X is a partition of X such that each part consists of integers consecutive in X. A segmentation σ of a source sentence S is a consecutive partition of $\{1, ..., |S|\}$. A part of σ, i.e., a segment, is intuitively interpreted as a phrase in S. In the graph representation G of (S, T) and A, a segmentation σ of S is realised by the existence of edges between consecutive source-type vertices whose labels, i.e., word positions in S, appear in the same segment of σ. The same argument holds for a target sentence and its words; a target segmentation is denoted by τ.

Clearly, there are $2^{|S|-1}$ possible ways to segment S and, given a fixed alignment matrix A, the number of all possible graphs that can be constructed is thus $2^{|S|+|T|-2}$. The bipartite graph of Section 3 is just one possible configuration, namely the one in which each segment of σ consists of exactly one word, and similarly for τ. We denote this segmentation pair by (σ_0, τ_0).

We now turn to extracting consistent pairs in this general setting from all possible segmentations (σ, τ) for a sentence pair (S, T) and a fixed alignment matrix A. As in Section 3, we construct graphs $G^{\sigma,\tau}$, associated sets $C_i^{\sigma,\tau}$, for all i, $1 \leq i \leq k^{\sigma,\tau}$, and $C^{\sigma,\tau}$, for all (σ, τ). Consistent pairs are extracted in lieu of eq. 4, i.e.,

$$P^{\sigma,\tau}(S,T) = \bigcup_{i=1}^{k^{\sigma,\tau}} \{ (S_H, T_H) : H \in C_i^{\sigma,\tau}, \\ S_H \subseteq S,\ T_H \subseteq T \}, \quad (9)$$

and it is trivial to see that

$$\{(S,T)\} \subseteq P^{\sigma,\tau}(S,T) \subseteq P(S,T), \quad (10)$$

for all (σ, τ). Note that $P(S,T) = P^{\sigma_0,\tau_0}(S,T)$ and, depending on the details of A, it is possible for other pairs (σ, τ) to attain equality. Moreover, each consistent pair in $P(S,T)$ can be be extracted from a member of at least one $C^{\sigma,\tau}$.

We focus on the sets $C_1^{\sigma,\tau}$, i.e., the components of $G^{\sigma,\tau}$, for all (σ, τ). In particular, we are interested in the relation between $P(S,T)$ and $C_1^{\sigma,\tau}$, for all (σ, τ). Each consistent $H \in C^{\sigma_0,\tau_0}$ can be converted into a single component by appropriately forming edges between consecutive source-type vertices and/or between consecutive target-type vertices. The resulting component will evidently be a member of $C_1^{\sigma,\tau}$, for some (σ, τ). It

[1] See Appendix for definitions.

is important to note that the conversion of a consistent $H \in C^{\sigma_0,\tau_0}$ into a single component need not be unique; see Figure 2 for a counterexample. Since (a) such conversions are possible for all consistent $H \in C^{\sigma_0,\tau_0}$ and (b) $P(S,T) = P^{\sigma_0,\tau_0}(S,T)$, it can be deduced that all possible consistent pairs can be traced in the sets $C_1^{\sigma,\tau}$, for all (σ, τ). In other words, we have:

$$P(S,T) = \bigcup_{\sigma,\tau} \{ (S_H,\, T_H) : \, H \in C_1^{\sigma,\tau}, \\ S_H \subseteq S,\, T_H \subseteq T \}. \quad (11)$$

The above equation says that by taking sentence segmentations into account, we can recover all possible consistent pairs, by inspecting only the components of the underlying graphs.

It would be interesting to investigate the relation between measure spaces $(C_1^{\sigma,\tau}, C^{\sigma,\tau}, f^{\sigma,\tau})$ and different configurations for A. We leave that for future work and focus on the advantages provided by eq. 11.

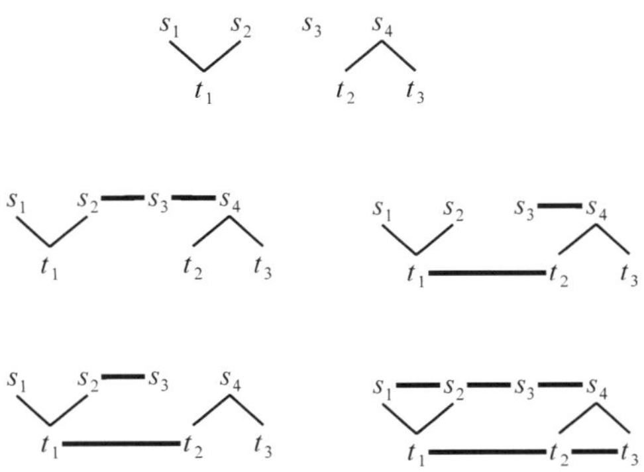

Figure 2: A graph with three components (top), and four possible conversions into a single component by forming edges between contiguous words.

5 Towards a phrase-level model that respects consistency

The aim of this section is to exploit the relation established in eq. 11 between consistent pairs and components of segmented sentence pairs. It was also shown in Section 2 that the computation of the translation models is inappropriate to describe the underlying structure. We thus suggest a phrase-based generative model in the spirit of the IBM word-based models, which is compatible with the construction of the previous sections.

5.1 Hidden variables

All definitions from the previous sections are carried over, and we introduce a new quantity that is associated with components. Let $G^{\sigma,\tau}$ and $C_1^{\sigma,\tau}$, for some (σ, τ) be as in Section 4, then the set K is defined as follows: Each member of K is a pair of (source, target) sets of segments that corresponds to the pair of (source, target) vertices of a consistent member of $C_1^{\sigma,\tau}$. In other words, K is a bisegmentation of a pair of segmented sentences that respects consistency.

Figure 3 shows three possible ways to construct consistent graphs from $(S,T) = (s_1^4, t_1^6)$, $\sigma = \{\{1,2\},\{3\},\{4\}\} \equiv \{x_1, x_2, x_3\}$ and $\tau = \{\{1\},\{2,3,4\},\{5\},\{6\}\} \equiv \{y_1, y_2, y_3, y_4\}$. In each case the exact alignment information is unknown and we have:

$$\text{(a) } K = \Big\{ \big(\{x_1\},\, \{y_1\}\big),\, \big(\{x_2\},\, \{y_2\}\big), \\ \big(\{x_3\},\, \{y_3, y_4\}\big) \Big\}.$$

$$\text{(b) } K = \Big\{ \big(\{x_1, x_2\},\, \{y_1, y_2, y_3\}\big), \\ \big(\{x_3\}, \{y_4\}\big) \Big\}.$$

$$\text{(c) } K = \Big\{ \big(\{x_1\},\, \{y_3, y_4\}\big), \\ \big(\{x_2,\, x_3\},\, \{y_1, y_2\}\big) \Big\}.$$

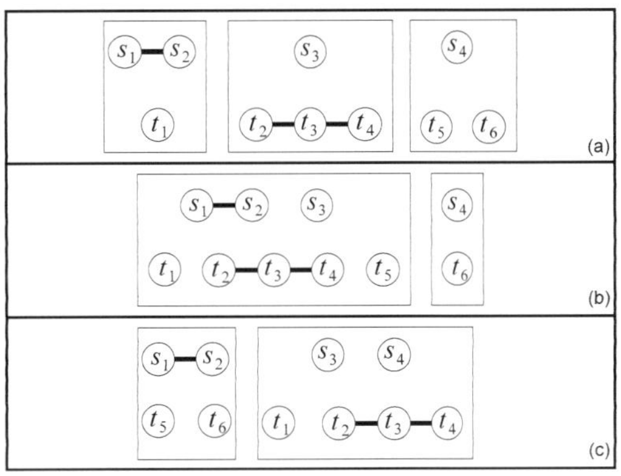

Figure 3: Three possible ways to construct consistent graphs for (s_1^4, t_1^6) and a given segmentation pair. Exact word alignment information is unknown.

In the proposed phrase-level generative model the random variables whose instances are σ, τ and

K are hidden variables. As with the IBM models, they are associated with the positions of words in a sentence, rather than the words themselves. Alignment information is implicitly identified via the consistent bisegmentation K.

Suppose we have a corpus that consists of pairs of parallel sentences (S, T), and let $f_{S,T}$ denote the occurrence count of (S, T) in the corpus. Also, let $l_S = |S|$ and $l_T = |T|$. The aim is to maximize the corpus log-likelihood function

$$\begin{aligned} \ell &= \sum_{S,T} f_{S,T} \log p_\theta(T|S) \\ &= \sum_{S,T} f_{S,T} \log \sum_{\sigma,\tau,K} p_\theta(T, \sigma, \tau, K|S), \end{aligned} \quad (12)$$

where σ, τ and K are hidden variables parameterized by a vector θ of unknown weights, whose values are to be determined. The expectation maximization algorithm (Dempster et al., 1977) suggests that an iterative application of

$$\theta_{n+1} = \arg\max_\theta \sum_{S,T} f_{S,T} \sum_{\sigma,\tau,K} p_{\theta_n}(\sigma, \tau, K|S, T) \times \log p_\theta(T, \sigma, \tau, K|S), \quad (13)$$

provides a good approximation for the maximum value of ℓ. As with the IBM models we seek probability mass functions (PMFs) of the form

$$p_\theta(T, \sigma, \tau, K|S) = p_\theta(l_T|S) p_\theta(\sigma, \tau, K|l_T, S) \times p_\theta(T|\sigma, \tau, K, l_T, S), \quad (14)$$

and decompose further as

$$p_\theta(\sigma, \tau, K|l_T, S) = p_\theta(\sigma, \tau|l_T, S) p_\theta(K|\sigma, \tau, l_T, S) \quad (15)$$

A further simplification of $p_\theta(\sigma, \tau|l_T, S) = p_\theta(\sigma|S) p_\theta(\tau|l_T)$ may not be desirable, but will help us understand the relation between θ and the PMFs. In particular, we give a formal description of $p_\theta(\sigma|S)$ and then explain that $p_\theta(K|\sigma, \tau, l_T, S)$ and $p_\theta(T|\sigma, \tau, K, l_T, S)$ can be computed in a similar way.

5.2 Constrained, biased sampling without replacement

The probability of a segmentation given a sentence can be realised in two different ways. We first provide a descriptive approach which is more intuitive, and we use the sentence $S = s_1^4$ as an example whenever necessary. The set of all possible segments of S is denoted by $seg(S)$ and trivially $|seg(S)| = |S|\big(|S| + 1\big)/2$. Each segment $x \in seg(S)$ has a nonnegative weight $\theta(x|l_S)$ such that

$$\sum_{x \in seg(S)} \theta(x|l_S) = 1. \quad (16)$$

Suppose we have an urn that consists of $|seg(S)|$ weighted balls; each ball corresponds to a segment of S. We sample without replacement with the aim of collecting enough balls to form a segmentation of S. When drawing a ball x we simultaneously remove from the urn all other balls x' such that $x \cap x' \neq \emptyset$. We stop when the urn is empty. In our example, let the urn contain 10 balls and suppose that the first draw is $\{1, 2\}$. In the next draw, we have to choose from $\{3\}$, $\{4\}$ and $\{3, 4\}$ only, since all other balls contain a '1' and/or a '2' and are thus removed. The sequence of draws that leads to a segmentation is thus a path in a decision tree. Since σ is a set, there are $|\sigma|!$ different paths that lead to its formation. The set of all possible segmentations, in all possible ways that each segmentation can be formed, is encoded by the collection of all such decision trees.

The second realisation, which is based on the notions of cliques and neighborhoods, is more constructive and will give rise to the desired PMF. A *clique* in a graph is a subset U of the vertex set such that for every two vertices $u, v \in U$, there exists an edge connecting u and v. For any vertex u in a graph, the *neighborhood* of u is defined as the set $N(u) = \{v : \{u, v\} \text{ is an edge}\}$. A *maximal clique* is a clique U that is not a subset of a larger clique: For each $u \in U$ and for each $v \in N(u)$ the set $U \cup \{v\}$ is not a clique.

Let $\mathcal{G}$ be the graph whose vertices are all segments of S and whose edges satisfy the condition that any two vertices x and x' form an edge iff $x \cap x' = \emptyset$; see Figure 4 for an example. $\mathcal{G}$ essentially provides a compact representation of the decision trees discussed above.

It is not difficult to see that a maximal clique also forms a segmentation. Moreover, the set of all maximal cliques in $\mathcal{G}$ is exactly the set of all possible segmentations for S. Thus, $p_\theta(\sigma|S)$ should satisfy

$$p_\theta(\sigma|S) = 0, \text{ if } \sigma \text{ is not a clique in } \mathcal{G}, \quad (17)$$

and

$$\sum_\sigma p_\theta(\sigma|S) = 1, \quad (18)$$

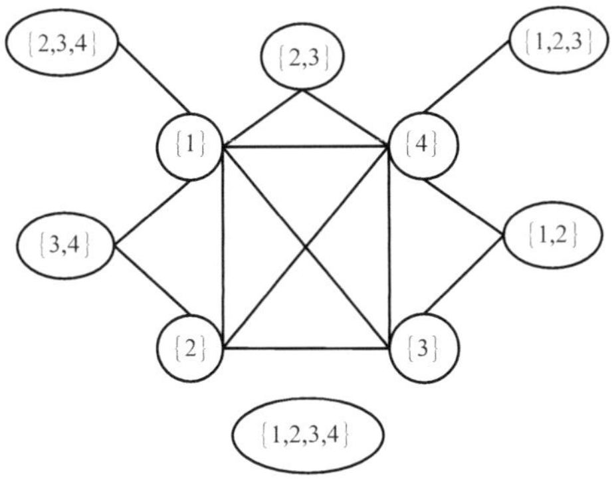

Figure 4: The graph whose vertices are the segments of s_1^4 and whose edges are formed by non-overlapping vertices.

where the sum is over all maximal cliques in $\mathcal{G}$. In our example $p_\theta(\{\ \{1\}, \{1,2\}\ \}|S) = 0$, because there is no edge connecting segments $\{1\}$ and $\{1,2\}$ so they are not part of any clique.

In order to derive an explicit formula for $p_\theta(\sigma|S)$ we focus on a particular type of paths in $\mathcal{G}$. A path is called *clique-preserving*, if every vertex in the path belongs to the same clique. Our construction should be such that each clique-preserving path has positive probability of occurring, and all other paths should have probability 0. We proceed with calculating probabilities of clique-preserving paths based on the structure of $\mathcal{G}$ and the constraint of eq. 16.

The probability $p_\theta(\sigma|S)$ can be viewed as the probability of generating all clique-preserving paths on the maximal clique σ in $\mathcal{G}$. Since σ is a clique, there are $|\sigma|!$ possible paths that span its vertices. Let $\sigma = \{x_1, ..., x_{|\sigma|}\}$, and let π denote a permutation of $\{1, ..., |\sigma|\}$. We are interested in computing the probability $q_\theta(x_{\pi(1)}, ..., x_{\pi(|\sigma|)})$ of generating a clique-preserving path $x_{\pi(1)}, ..., x_{\pi(|\sigma|)}$ in $\mathcal{G}$. Thus,

$$\begin{aligned} p_\theta(\sigma|S) &= p_\theta(\{x_1, ..., x_{|\sigma|}\}|S) \\ &= \sum_\pi q_\theta(x_{\pi(1)}, ..., x_{\pi(|\sigma|)}) \\ &= \sum_\pi q_\theta(x_{\pi(1)})\, q_\theta(x_{\pi(2)}|x_{\pi(1)}) \times ... \\ &... \times q_\theta(x_{\pi(|\sigma|)}|x_{\pi(1)}, ..., x_{\pi(|\sigma|-1)}). \end{aligned} \tag{19}$$

The probabilities $q_\theta(\cdot)$ can be explicitly calculated by taking into account the following observation. A clique-preserving path on a clique σ can be realised as a sequence of vertices $x_{\pi(1)}, ..., x_{\pi(i)}, ..., x_{\pi(|\sigma|)}$ with the following constraint: If at step $i-1$ of the path we are at vertex $x_{\pi(i-1)}$, then the next vertex $x_{\pi(i)}$ should be a neighbor of all of $x_{\pi(1)}, ..., x_{\pi(i-1)}$. In other words we must have

$$x_{\pi(i)} \in N_{\pi,i} \equiv \bigcap_{l=1}^{i-1} N(x_{\pi(l)}). \tag{20}$$

Thus, the probability of choosing $x_{\pi(i)}$ as the next vertex of the path is given by

$$q_\theta(x_{\pi(i)}|x_{\pi(1)}, ..., x_{\pi(i-1)}) = \frac{\theta(x_{\pi(i)}|l_S)}{\sum_{x \in N_{\pi,i}} \theta(x|l_S)}, \tag{21}$$

if $x_{\pi(i)} \in N_{\pi,i}$ and 0, otherwise. When choosing the first vertex of the path (the root in the decision tree) we have $N_{\pi,1} = seg(S)$, which gives $q_\theta(x_{\pi(1)}) = \theta(x_{\pi(1)}|l_S)$, as required. Therefore eq. 19 can be written compactly as

$$p_\theta(\sigma|S) = \left(\prod_{i=1}^{|\sigma|} \theta(x_i|l_S)\right) \sum_\pi \frac{1}{Q_\theta(\sigma, \pi; S)}, \tag{22}$$

where

$$Q_\theta(\sigma, \pi; S) = \prod_{i=1}^{|\sigma|} \sum_{x \in N_{\pi,i}} \theta(x|l_S)\ . \tag{23}$$

The construction above can be generalized in order to derive a PMF for any random variable whose values are partitions of a set. Indeed, by allowing the vertices of $\mathcal{G}$ to be a subset of a power set, and keeping the condition of edge formation the same, probabilities of clique-preserving paths can be calculated in the same way. Figure 5 shows the graph $\mathcal{G}$ that represents all possible instances of K with $(S,T) = (s_1^4, t_1^5)$, $\sigma = \{\{1,2\},\{3\},\{4\}\}$ and $\tau = \{\{1\},\{2,3,4\},\{5\}\}$. Again each maximal clique is a possible consistent bisegmentation.

In order for this model to be complete, one should solve the maximization step of eq. 13 and calculate the posterior $p_{\theta_n}(\sigma, \tau, K|S,T)$. We are not bereft of hope, as relevant techniques have been developed (see Section 6).

6 Related Work

To our knowledge, this is the first attempt to investigate formal motivations behind the consistency method.

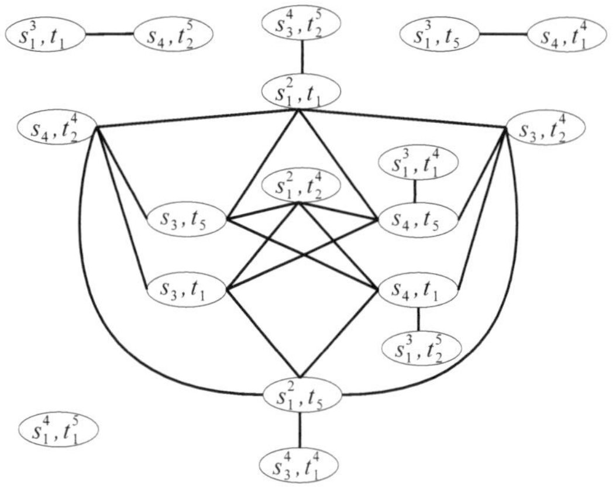

Figure 5: Similar to Figure 4 but for consistent bisegmentations with $(S, T) = (s_1^4, t_1^5)$ and a given segmentation pair (see text). For clarity, we show the phrases that are formed from joining contiguous segments in each pair, rather than the segments themselves.

Several phrase-level generative models have been proposed, almost all relying on multinomial distributions for the phrase alignments (Marcu and Wong, 2002; Zhang et al., 2003; Deng and Byrne 2005; DeNero et al., 2006; Birch et al., 2006). This is a consequence of treating alignments as functions rather than partitions.

Word alignment and phrase extraction via Inversion Transduction Grammars (Wu, 1997), is a linguistically motivated method that relies on simultaneous parsing of source and target sentences (DeNero and Klein, 2010; Cherry and Lin 2007; Neubig et al., 2012).

The partition probabilities we introduced in Section 5.2 share the same tree structure discussed in (Dennis III, 1991), which has found applications in Information Retrieval (Haffari and Teh, 2009).

7 Conclusions

We have identified the relation between consistency and components of graphs that represent word-aligned sentence pairs. We showed that phrase pairs of interest to SMT form a sigma-algebra generated by components of such graphs, but the existing occurrence-counting statistics are inadequate to describe this structure. A generalization of our construction via sentence segmentations lead to a realisation of random partitions as cases of constrained, biased sampling without replacement. As a consequence, we derived an exact formula for the probability of a segmentation of a sentence.

Appendix: Measure Space

The following standard definitions can be found in, e.g., (Feller, 1971). Let X be a set. A collection B of subsets of X is called a *sigma-algebra* if the following conditions hold:

1. $\emptyset \in B$.
2. If E is in B, then so is its complement $X \setminus E$.
3. If $\{E_i\}$ is a countable collection of sets in B, then so is their union $\cup_i E_i$.

Condition 1 guarantees that B is non-empty and Conditions 2 and 3 say that B is closed under complementation and countable unions respectively. The pair (X, B) is called a *measurable space*.

A function $f : B \to [0, \infty)$ is called a *measure* if the following conditions hold:

1. $f(\emptyset) = 0$.
2. If $\{E_i\}$ is a countable collection of pairwise disjoint sets in B, then

$$f(\cup_i E_i) = \sum_i f(E_i).$$

Condition 2 is known as *sigma-additivity*. The triple (X, B, f) is called a *measure space*.

Acknowledgments

This research was supported by the European Union's ICT Policy Support Programme as part of the Competitiveness and Innovation Framework Programme, CIP ICT-PSP under grant agreement nr 250430 (GALATEAS) and by the EC funded project CoSyne (FP7-ICT-4-24853).

References

Alexandra Birch, Chris Callison-Burch, Miles Osborne and Philipp Koehn. 2006. Constraining the Phrase-Based, Joint Probability Statistical Translation Model. In *Proc. of the Workshop on Statistical Machine Translation*, pages 154–157.

Peter F. Brown, Stephen A. Della Pietra, Vincent J. Della Pietra, and Robert L. Mercer. 1993. The Mathematics of Statistical Machine Translation. *Computational Linguistics*, vol.19(2), pages 263–312.

Colin Cherry and Dekang Lin. 2007. Inversion Transduction Grammar for Joint Phrasal Translation Modeling. In *Proc. of SSST, NAACL-HLT / AMTA Workshop on Syntax and Structure in Statistical Translation*, pages 17–24.

A.P. Dempster, N.M. Laird and D.B. Rubin. 1977. Maximum Likelihood from Incomplete Data via the EM Algorithm. *Journal of the Royal Statistical Society*, Series B (Methodological) 39(1), pages 1–38.

John DeNero, Dan Gillick, James Zhang and Dan Klein. 2006. Why Generative Phrase Models Underperform Surface Heuristics. In *Proc. of the Workshop on Statistical Machine Translation*, pages 31–38.

John DeNero and Dan Klein. 2010. Discriminative Modeling of Extraction Sets for Machine Translation. In *Proc. of the Association for Computational Linguistics (ACL)*, pages 1453–1463.

Yonggang Deng and William Byrne. 2005. HMM Word and Phrase Alignment for Statistical Machine Translation. In *Proc. of the Conference on Empirical Methods in Natural Language Processing and Human Language Technology (HLT-EMNLP)*, pages 169–176.

Samuel Y. Dennis III. 1991. On the Hyper-Dirichlet Type 1 and Hyper-Liouville Distributions. *Communications in Statistics - Theory and Methods*, 20(12), pages 4069–4081.

William Feller. 1971. *An Introduction to Probability Theory and its Applications, Volume II.* John Wiley, New York.

Gholamreza Haffari and Yee Whye Teh. 2009. Hierarchical Dirichlet Trees for Information Retrieval. In *Proc. of the Conference of the North American Chapter of the Association for Computational Linguistics on Human Language Technology (HLT-NAACL)*, pages 173–181.

Frank Harary. 1969. *Graph Theory*. Addison–Wesley, Reading, MA.

Philipp Koehn, Franz Josef Och, and Daniel Marcu. 2003. Statistical Phrase-Based Translation. In *Proc. of the Conference of the North American Chapter of the Association for Computational Linguistics on Human Language Technology (HLT-NAACL)*, pages 48–54.

Philipp Koehn. 2009. *Statistical Machine Translation*. Cambridge University Press, Cambridge, UK.

Daniel Marcu and William Wong. 2002. A Phrase-Based, Joint Probability Model for Statistical Machine Translation. In *Proc. of the Conference on Empirical Methods in Natural Language Processing (EMNLP)*, pages 133–139.

Graham Neubig, Taro Watanabe, Eiichiro Sumita, Shinsuke Mori and Tatsuya Kawahara. 2012. Joint Phrase Alignment and Extraction for Statistical Machine Translation. *Journal of Information Processing*, vol. 20(2), pages 512–523.

Franz J. Och, Christoph Tillmann, and Hermann Ney. 1999. Improved Alignment Models for Statistical Machine Translation. In *Proc. of the Joint Conference of Empirical Methods in Natural Language Processing and Very Large Corpora (EMNLP-VLC)*, pages 20–28.

Dekai Wu. 1997. Stochastic Inversion Transduction Grammars and Bilingual Parsing of Parallel Corpora. *Computational Linguistics*, 23, pages 377–404.

Kenji Yamada and Kevin Knight. 2001. A syntax-based statistical translation model. In *Proc. of the Association for Computational Linguistics (ACL)*, pages 523–530.

Ying Zhang, Stephan Vogel and Alex Waibel. 2003. Integrated Phrase Segmentation and Alignment Algorithm for Statistical Machine Translation. In *Proc. of the International Conference on Natural Language Processing and Knowledge Engineering (NLP-KE)*.

Grammars and Topic Models

Mark Johnson
Centre for Language Sciences and Dept. of Computing
Macquarie University
Sydney, Australia
Mark.Johnson@MQ.edu.au

1 Abstract

Context-free grammars have been a cornerstone of theoretical computer science and computational linguistics since their inception over half a century ago. Topic models are a newer development in machine learning that play an important role in document analysis and information retrieval. It turns out there is a surprising connection between the two that suggests novel ways of extending both grammars and topic models. After explaining this connection, I go on to describe extensions which identify topical multiword collocations and automatically learn the internal structure of named-entity phrases.

The adaptor grammar framework is a non-parametric extension of probabilistic context-free grammars (Johnson et al., 2007), which was initially intended to allow fast prototyping of models of unsupervised language acquisition (Johnson, 2008), but it has been shown to have applications in text data mining and information retrieval as well (Johnson and Demuth, 2010; Hardisty et al., 2010). We'll see how learning the referents of words (Johnson et al., 2010) and learning the roles of social cues in language acquisition (Johnson et al., 2012) can be viewed as a kind of topic modelling problem that can be reduced to a grammatical inference problem using the techniques described in this talk.

2 About the Speaker

Mark Johnson is a Professor of Language Science (CORE) in the Department of Computing at Macquarie University in Sydney, Australia. He was awarded a BSc (Hons) in 1979 from the University of Sydney, an MA in 1984 from the University of California, San Diego and a PhD in 1987 from Stanford University. He held a postdoctoral fellowship at MIT from 1987 until 1988, and has been a visiting researcher at the University of Stuttgart, the Xerox Research Centre in Grenoble, CSAIL at MIT and the Natural Language group at Microsoft Research. He has worked on a wide range of topics in computational linguistics, but his main research areas are computational models of language acquisition, and parsing and its applications to text and speech processing. He was President of the Association for Computational Linguistics in 2003 and is Vice-President elect of EMNLP, and was a professor from 1989 until 2009 in the Departments of Cognitive and Linguistic Sciences and Computer Science at Brown University.

References

Eric A. Hardisty, Jordan Boyd-Graber, and Philip Resnik. 2010. Modeling perspective using adaptor grammars. In *Proceedings of the 2010 Conference on Empirical Methods in Natural Language Processing*, pages 284–292, Stroudsburg, PA, USA. Association for Computational Linguistics.

Mark Johnson and Katherine Demuth. 2010. Unsupervised phonemic Chinese word segmentation using Adaptor Grammars. In *Proceedings of the 23rd International Conference on Computational Linguistics (Coling 2010)*, pages 528–536, Beijing, China, August. Coling 2010 Organizing Committee.

Mark Johnson, Thomas L. Griffiths, and Sharon Goldwater. 2007. Adaptor Grammars: A framework for specifying compositional nonparametric Bayesian models. In B. Schölkopf, J. Platt, and T. Hoffman, editors, *Advances in Neural Information Processing Systems 19*, pages 641–648. MIT Press, Cambridge, MA.

Mark Johnson, Katherine Demuth, Michael Frank, and Bevan Jones. 2010. Synergies in learning words and their referents. In J. Lafferty, C. K. I. Williams, J. Shawe-Taylor, R.S. Zemel, and A. Culotta, editors, *Advances in Neural Information Processing Systems 23*, pages 1018–1026.

Mark Johnson, Katherine Demuth, and Michael Frank. 2012. Exploiting social information in grounded language learning via grammatical reduction. In

Proceedings of the 50th Annual Meeting of the Association for Computational Linguistics, pages 883–891, Jeju Island, Korea, July. Association for Computational Linguistics.

Mark Johnson. 2008. Using Adaptor Grammars to identify synergies in the unsupervised acquisition of linguistic structure. In *Proceedings of the 46th Annual Meeting of the Association of Computational Linguistics*, pages 398–406, Columbus, Ohio. Association for Computational Linguistics.

Author Index

Association for Computational Linguistics
209 N. Eighth Street
Stroudsburg, Pennsylvania 18360

CURRAN ASSOCIATES INC
proceedings.com